Artur Torres Araújo

Problematizing thermodynamic concepts

AF671205

Artur Torres Araújo

Problematizing thermodynamic concepts

Cold influencing contexts

ScienciaScripts

Imprint
Any brand names and product names mentioned in this book are subject to trademark, brand or patent protection and are trademarks or registered trademarks of their respective holders. The use of brand names, product names, common names, trade names, product descriptions etc. even without a particular marking in this work is in no way to be construed to mean that such names may be regarded as unrestricted in respect of trademark and brand protection legislation and could thus be used by anyone.

Cover image: www.ingimage.com

This book is a translation from the original published under ISBN 978-3-330-77025-6.

Publisher:
Sciencia Scripts
is a trademark of
Dodo Books Indian Ocean Ltd. and OmniScriptum S.R.L publishing group

120 High Road, East Finchley, London, N2 9ED, United Kingdom
Str. Armeneasca 28/1, office 1, Chisinau MD-2012, Republic of Moldova, Europe
Managing Directors: Ieva Konstantinova, Victoria Ursu
info@omniscriptum.com

Printed at: see last page
ISBN: 978-620-8-63063-8

Copyright © Artur Torres Araújo
Copyright © 2025 Dodo Books Indian Ocean Ltd. and OmniScriptum S.R.L publishing group

SUMMARY

SUMMARY

This research was carried out at the Imaculada Conceiçao State School, in the city of Cabedelo, with 40 high school students. To carry out this work, two learning situations (LS) were developed, focusing on the *"initial problematization"*, the first of three stages of a dynamic called *pedagogical moments* proposed by Delizoicov. A teaching situation (SE) was developed to guide the application of the LS. The TS were developed with the students in the form of texts containing the history of science and questions in which the students had to debate, argue and develop an answer to questions about the text and everyday life in relation to the concepts of heat and temperature. To complete the practice, a pre-test was given with multiple choice questions on the concepts of heat and temperature. The students' answers were analyzed using the *conceptual profiles* approach proposed by Mortimer. The ideas expressed by the students were compared with current scientific ideas in order to diagnose conceptual cognitive barriers. In the analysis of the students' answers to the SA, 25 conceptual cognitive barriers were identified, 10 of which were in the category related to the concepts of cold, heat and temperature, and were classified as belonging to the animist, substantialist and realist zones of the conceptual profile. In the pre-test, 9 were identified, of which 2 had already been identified in the students' answers regarding AS. These 9 conceptual cognitive barriers were classified as belonging to the substantialist and realist zones of the conceptual profile. These conceptual cognitive barriers should be taken into account by the teacher during the development of lessons and, with this, the teacher should seek to develop practices aimed at overcoming them, making students aware of the need to build new (scientific) knowledge that can better explain phenomena that occur on a daily basis, and making them rethink the world in which they live.

KEYWORDS: Heat and Temperature. Pedagogical Moments. Conceptual Profiles.

INTRODUCTION

In order to carry out this work, learning situations were developed, focusing on the "initial problematization", considered to be the first of three stages in a dynamic called pedagogical moments (Delizoicov 1991, 2008).

The *initial problematization* is a diagnostic phase in which the researcher will be able to understand the students' previous conceptions of a topic and whether or not these conceptions are far removed from scientific conceptions. At this point, the topic to be worked on in the classroom should not be presented directly, but the student should be put in contact with the subject and try to extract information about their understanding of the subject.

In order to construct the initial problematization, two learning situations (LS) were developed and applied to students in the form of texts and questions dealing with the history of science.

A teaching situation (SE) was developed to guide the teacher's application of SA in the classroom.

There is also another dynamic called the *"study situation"* which is also divided into three stages and is very similar to the dynamics of the pedagogical moments, one of the main differences being that the study situation is based on Vygotsky's ideas.

In this work, we opted for the pedagogical moments technique, because in an article by Gehlen, Maldaner and Delizoicov (2012) discussing the process of developing the two aforementioned techniques in the classroom, it was observed that in the study situation, the themes addressed are not evident, and this is where it differs from the Freirean theme where it is assumed that thematic investigation is the dynamic for acquiring the themes that will be developed.

In Freire's proposal, the problem will organize the entire didactic-pedagogical process in the thematic investigation, in which *"problematization consists of addressing certain problems which are manifestations of local contradictions"* (FREIRE, 2005). Therefore, the teaching practice in the *initial problematization* is not to introduce or expose the word that is directly linked to the scientific concept, which, unlike in the study situation, is paramount.

The *initial problematization* should make the student aware of the need for new (scientific) knowledge that can contribute to the understanding of a problem, and prepare the introduction of the scientific concept for the second pedagogical moment.

Intuitively, teachers carry out the second (organization of knowledge) and third pedagogical moments (application of knowledge). In short, the second pedagogical moment is where the information will be systematized with the aim of transforming it into knowledge and in the third pedagogical moment this constructed knowledge will be used to answer the questions of the first pedagogical moment and new questions or problems that have arisen.

Therefore, we believe that the initial problematization is an important moment that is often neglected by teachers and does not have its deserved place in teaching practices in both basic and higher education. This moment provides valuable epistemological and ontological information for teachers and their practice, such as the concept of the world that students bring with them and the possible *"conceptual cognitive barriers"* that need to be overcome in order to interpret certain phenomena in a new way.

With the introduction of the SA problematizers, 25 conceptual cognitive barriers were diagnosed, 10 of which belonged to the category of cold, heat and temperature concepts, and were classified as belonging to the animist, substantialist and realist zones of the conceptual profile of heat. By identifying these conceptual cognitive barriers, the teacher will be able to propose and develop practices (different or not from the practices proposed by the textbook adopted) aimed at overcoming these barriers.

With this in mind, the objectives of this work were:

Prepare problematizing learning situations on the topics of heat and temperature;

Use the documentary "Absolute zero: the conquest of the cold" as a source of content;

Prepare a teaching situation to guide the application of learning situations in the classroom;

Applying learning situations in the classroom;

Prepare a pre-test on scientific knowledge of heat and temperature;

Analyze the results of the pre-test, the application of the teaching situation and the application of the learning situations in the classroom;

Identify conceptual cognitive barriers.

With the application of problematizing AS, it was hoped to awaken a participatory and dynamic interest in the students, since it would be a different practice from those found in most textbooks. However, almost all the students didn't like the practice, as they would have to read, reflect, create and discuss answers.

When we presented this problematizing proposal to some secondary school teachers, we got the following response: *"Paulo Freire's education is very beautiful, but if Paulo Freire were still alive I'd like to see him come here and teach these students today to see if he would still have the same ideas."*

Adhering to these problematizing proposals is very complicated, as it requires a lot of dedication on the part of the teacher to produce situations to diagnose conceptual cognitive barriers, to make analyses of the students' ideas and to adapt the school curriculum (creating new practices or content) to meet the students' conceptual needs. Such practices require a lot of time reading and preparing teaching material, which is why most primary school teachers don't want to leave their comfort zone, and the proposal becomes more attractive if they follow the textbook.

CHAPTER I: THEORETICAL BACKGROUND

1.1. Teaching heat and temperature in literature

Great efforts have been made to teach chemistry more effectively and in a way that makes sense in students' lives. However, in this scenario, there is a lack of connection between the ideas and languages created by science and the ideas and languages created by popular knowledge, which does not allow students to reflect on their knowledge and compare it with the new things they learn in chemistry classes.

These and other actions make the science that is taught in schools uninteresting, fruitless and without logic in students' daily lives. Failure in the natural sciences expresses a false notion that students are incapable of understanding the world around them and that science is for the intellectually gifted few (TUNES, 1995).

Teachers and scientists have contributed to improving science teaching. However, science is still seen as a set of immutable truths, of conceptual structures frozen in time. It often has no relation to the historical, social and technological contexts in which science is constructed and applied (LABURU, 1995; HELLER, 1989 and 1991).

In the study of thermodynamics, some concepts such as energy, heat and temperature are widely used and we have become accustomed to using them in our daily lives. These concepts do not have the same meaning in science as they do in popular language. These facts have been the subject of discussions about difficulties in teaching chemistry, since teachers work with more complex concepts, such as heat of reaction, Hess's Law, etc., often without reviewing more basic concepts (MORTIMER and AMARAL, 1998).

The concept of heat has been much discussed and has been the subject of various studies in science teaching, which have dealt with the history of science, students' spontaneous conceptions, suggestions for teaching strategies, difficulties in the teaching and learning process and others. Among these studies, Albert (1978); Brook, Briggs, Bell and Driver (1984); Cervantes (1987); Erickson (1985); Silva (1995); Barbosa Lima and Barros (1997); Mortimer and Amaral (1998) stand out.

More recent work has addressed the difficulties students have in interpreting the concepts of heat and temperature, showing that this is not a new problem for chemistry teaching researchers (GRINGS, CABALLERO and MOREIRA, 2006; 2007 and 2008). Some researchers have developed proposals and teaching materials to try to help students learn these concepts. In the literature, there is some research into the development of didactic proposals for thermodynamics, with an emphasis on addressing the concepts of heat and temperature.

Carvalho and colleagues (1999) developed a book based on inquiry teaching, presenting a set of inquiry activities on topics such as temperature measurements, insulation and conductors, expansion, calorimetry, phase changes, among others. The text not only explores educational proposals with an investigative basis, but also comments on their application in the classroom and the attitudes of teachers and students. The proposal is based on investigative reading of historical texts and investigative experimental demonstrations, open laboratories, open questions and problems. At the end of the book, the authors propose a six-month course in thermal physics and suggestions for videos, films, programs and websites on the subject, in addition to the extensive literature entrusted with the work.

Boss and colleagues (2009) reported on an educational experiment in thermodynamics also based on inquiry teaching. Experiments have been proposed to measure temperature by touch and heat by conduction and convection. The authors used historical texts to conceptualize heat. The concept of temperature was discussed. The students also made microscopic models to represent the structure of matter, temperature and expansion.

Some dissertations are worth mentioning, such as those by Gonçalves (2005), Rafael (2007) and Silva Jr. (2007). Gonçalves (2005) presented a proposal based on educational technologies (such as videos, animations and interactive simulations of phenomena) as complementary activities to lessons. These materials were applied in a public school, with students in the 2nd year of secondary education. It also presents videotaped experiments on expansion, modes of heat propagation and thermal machines.

The work is based on David Ausubel's meaningful learning, which describes a series of animations he has produced on a wide range of thermal physics topics such as temperature, thermometric scales, kinetic energy, expansion, heat, thermal equilibrium, heat capacity, specific heat, solid conduction, the structure of materials, evaporation, the first law of thermodynamics, entropy and thermal machines.

Rafael's work (2007) was applied to a 2nd year high school class in a public school and investigates students' alternative conceptions of heat and temperature in order to provide an educational proposal that includes the following elements: reading and discussion of texts on the history of science, focusing on the evolution of the concepts of heat and temperature, thermal machines and the industrial revolution and its consequences; reading and discussion of current texts on the construction of the concepts of heat and temperature; and construction of a prototype thermal machine using low-cost materials.

Silva Jr. (2007) developed lesson plans for learning thermal physics content in a 2nd year high school class, using texts from the Internet as the main educational resources, with an interdisciplinary focus. Each lesson plan had the following structure: basic text, teaching levels, materials, lesson plan overview, time given, resources/materials, activities/procedures, lesson objectives, discussion questions, content summary, Internet links and references, interdisciplinary connections, National Curriculum Parameters. In turn, temperature, thermometric scales, expansion, heat and forms of transfer, pressure, thermal machines, sources and transformations of energy, power, performance and operation of motors were the physics topics covered in the lesson plans, which also suggested carrying out experiments.

1.2. The concepts of heat and temperature from 3 perspectives: everyday, historical and scientific

Temperature is one of the most well-known and quoted scientific concepts today. Every day people watch the weather on television, tablets and smartphones, indicating the temperature range for their region. We adjust the temperature of the air conditioning unit and check our body temperature to see if we are feverish and relate the temperature to the individual's state of health.

Temperature gives us information about how hot a body is in relation to another, it also acts as an indicator of the direction of the exchange of energy in the form of heat, and can be perceived in various ways.

Heat and temperature are subjects that are easy to understand because of their influence on everyday life. Every day people communicate using the concepts of heat and temperature and practically all of this communication is done using popular or common sense language which often disagrees with the language of scientific ideas, in this particular case the concepts of heat and temperature used by popular language are in disagreement with the understanding of these concepts by the scientific conception, however even scientists use popular language in their daily lives.

According to Silva, Neto and Carvalho (1988), it is common to use the concepts of heat and temperature wrongly from a scientific point of view, and to use them as synonyms or even as a property of something: "today is very hot", "he burnt his hand because the pan got hot", "the heat of the sun", "today it is 40°C hot", "the ceramic floor is very cold", among many others that we see on a daily basis. This demonstrates the spontaneous thinking of people, even those who are educated and in academia.

Even large electronic magazines, such as the famous "Fantàstico" on the Globo network, recently made these mistakes when they published on their website the G1 portal, in the edition of February 9, 2014, shown in figures 1 and 2, where it brings some news such as:

"In Sorocaba, Sao Paulo, a woman had to go to the fire station to cut off her wedding ring. Her fingers were so swollen from the almost 36°C heat that there was no way the ring would come off."

"And hot water helps to increase the heat in the earth",

"Even without rain, clouds are important for reducing the heat, not just because they act like an umbrella. The higher they are and the more ice crystals they contain, the cooler the environment. It's just that, high up, they act as a reflector, throwing the sun's light and heat back into the atmosphere." (G1/FANTÂSTICO, 2014).

Figura 1 - Headline in the electronic magazine Fantastico.

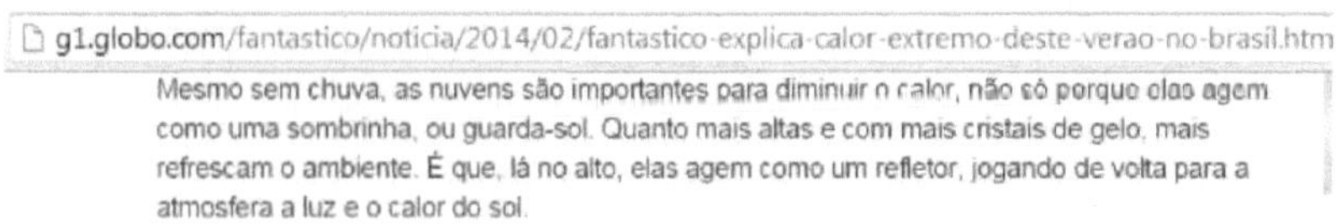

Figure 2 - Headline in the electronic magazine Fantàstico.

Another important electronic magazine also made a mistake when it published: "6 reasons that make a 50°C heat hell", in its publication of July 7, 2013 (EXAME, 2013).

Figure 3 - Exame magazine headline.

The most primitive ideas of heat are those born from the sensations of hot and cold, which are associated with the idea of hot, or the thermal sensation of warmth and were directly related to fire. This was used as a source of light or heat, and was also used as a weapon and source of energy for the transformation of materials, revolutionizing human life.

Man's fascination with fire, and therefore with heat, dates back to the dawn of civilization, when this resource was first used for heating and cooking. It was the cavemen who first tried to understand the mystery of heat. As he states, the notions related to heat are innate to *homo erectus* (approximately 700,000 years ago), when the dominance of fire as a source of heat began (SILVA, 2007).

In the history of science, it is possible to see some events that may have led to this confusion in the use of the concepts of heat and temperature. For Galen, a famous physician who lived between 129-200, the human body was a "*mixture*" of heat and cold, which determined, among other things, the patient's state of health.

For Combie (1985), the influences of Galen's work marked medicine, as his work was considered the main medical guide for a long time. When Galen's works were translated into Latin in the 11th and 12th centuries, the word mescla became equivalent to ***"tempera"*** (temperature). With this, the association of temperature as a measure of heat began.

Even at the beginning of the 17th century, there were two currents that tried to explain heat. The first associated heat with the idea of fluid, which began with ancient Greek thinkers who understood heat as a substance that inhabited matter and was absorbed or released by various processes. Among its defenders, the most famous was Lavoisier and his caloric theory, which understood heat to be a substance that he called *"caloric"* in his works, going so far as to put caloric on a list of substances. The second associated heat with the movement of particles in a body. This model is attributed to Francis Bacon, who in his works presented heat as vibrating movements of the particles of a body (HOPPE, 1928; SCHURMANN, 1946; BASSALO, 1991).

There was no consensus on which of the two was correct, perhaps because until then there was no need for a quantitative treatment of heat. During this period there seems to have been no concern with differentiating the concepts of heat and temperature (CINDRA and TEIXEIRA, 2004).

Proponents of heat as a substance created various ways of trying to measure the mass of heat, and in 1785 the doctor George Fordyce carried out famous experiments to determine whether frozen water varied in mass. Thompson and Rumford repeated these experiments and presented a paper concluding that all attempts to observe the effect of heat on mass were futile (BASSALO, 1992).

It is now accepted that matter is made up of atoms and molecules that have stored energy. Whatever their physical state, molecules are in constant movement, which characterizes the existence of kinetic energy in their stock. Potential energy is the result of interactions with neighboring molecules. Energy also exists in matter as a result of its mass. The internal energy found inside matter is the sum of all these types of energy (ATKINS and JONES, 2006).

Temperature, which in common sense is understood as the measure of heat in bodies, in science is the measure of the kinetic energy of the movement of atoms. This makes kinetic energy directly proportional to temperature. It is important to note that in the example of phase changes there is a variation in internal energy without a variation in temperature.

In order for energy to be transferred in the form of heat from one body to another, at least two bodies with different temperatures must interact. As a result, the warmer body (the one with the higher temperature) begins to release energy in the form of heat (the name given to energy in transit) to the colder body (the one with the lower temperature). As a result of this interaction, the temperature of the warmer body decreases (so its kinetic energy also decreases) and that of the colder body increases (so its kinetic energy also increases). The body with the higher temperature releases heat to the body with the lower temperature until the bodies have the same temperature, i.e. when they reach thermal equilibrium (MARQUES and ARAÙJO, 2009).

Mortimer and Machado (2013) argue that the sum of a body's energies is called internal energy, and when two bodies are in contact at different temperatures, the body with the higher temperature transfers part of its internal energy to the body with the lower temperature, and this process only ceases when the two bodies reach the same temperature (thermal equilibrium).

Heat is then understood as the energy that has been transferred to the body with the lowest temperature. No body has heat, but it does have internal energy, and when part of this energy is in transit it is called heat. As soon as the energy transferred in the form of heat reaches the body with the lowest temperature, it ceases to be heat and becomes part of that body's internal energy.

In this way, we can see the distance between the concepts of heat and temperature from the point of view of science and from the point of view of common sense, and that this distance can cause confusion when it comes to teaching current scientific concepts.

The history of science proves to be a powerful pedagogical tool, since the student has the opportunity to come into contact with the construction of concepts that great scientists have experienced over time, realizing the social, economic, political, cultural and geographical context of the people who have contributed significantly to the advancement of a given scientific concept, which can contribute to the student's conceptual development.

However, it's not enough just to combine history with science and think that you're working on the history of science. This pedagogical tool has to be used in such a way as to offer students the chance to understand how scientists arrived at a certain concept. Does the era in which a scientist lives influence his or her thinking? If Haber hadn't lived in a time of war, would he have developed his famous process for obtaining ammonia? Would Copernicus have defended heliocentrism if he had lived in a time when the Church wasn't so powerful? Would Lavoisier have been a tax collector in other times?

These contexts, when perceived, can promote a better dialogue between common sense and science, and the two begin to coexist in the individual's head (in the model called *"conceptual profiles"*, which will be discussed in the next section).

1.3 Concepts and conceptualizations

The most widely accepted idea of a "concept" today is that of a mental model or schema constructed by the learner to represent objects or events, which becomes relatively stable when possessed by an individual. When an individual model undergoes some kind of transformation, it is believed that a conceptual evolution has taken place. This process can signify student learning (MORTIMER, SCOTT and EL-HANI, 2011).

Conceptualization is a process of constructing concepts that originates in the social interaction between an individual and some external experience. Conceptual thinking is limited by the meanings of certain concepts mastered by individuals. Socialization processes are strong because they indicate the feeling that we "own" concepts in our minds.

From a Vygotskian perspective, conceptual thinking is a tendency or potentiality to remain in that sensation, when fully developed. As an emergent process, it tends to repeat apparently central characteristics, which allow us to use concepts repeatedly, and thus effectively think through concepts and communicate through the signs of language (VYGOTSKY, 1978).

The distinction between sense and meaning broadens the understanding of what is permanent in conceptualization. "*The meaning of a word is the aggregate of all the psychological facts that arise in our consciousness as a result of the word. Meaning is a dynamic, fluid and complex formation that has several zones that vary in their stability. In different contexts, the meaning of a word changes*" (VYGOTSKI, 2000). The social construction known as "meaning" makes subjectivity possible, and people have the possibility of sharing the meaning of a word, even if the meanings attributed to it differ. In this way, the word becomes the carrier of the concept.

These socio-culturally stable constructions are the basis for communication contracts (ROMMETVEIT, 1979). Learning a concept is directly linked to learning its meaning, passing on personal meanings to socially accepted meanings. The formation of meaning is a purely personal development. Each individual generates different meanings for the same word and can also vary the meanings generated depending on the context.

When we work as teachers in a classroom, we can see the diversity of thoughts and ways in which students see and describe the world around them. Should these ideas be disregarded? Or can we build meaningful knowledge with this heterogeneity of ideas and languages expressed in chemistry classes?

Mortimer (1994, 1995) considered the diversity of language and thought in science classrooms and proposed *"conceptual profiles"* as a way of modeling this heterogeneity of verbal thought in science classrooms. *Conceptual profiles* should represent models of the different ways of seeing and representing the world used by people to give meaning to their experience.

Conceptual profiles emerged as a complement to the conceptual change model of Posner et al. (1982), diverging strongly from a central idea that students should be led to break with their previous conceptions when learning science. This divergence brings conceptual profiles closer to William Cobern's conceptual constructivism (1996), which also proposes the coexistence of different ways of thinking and speaking as a result of learning science.

The idea of a *conceptual profile* can be used to describe the evolution and structuring of ideas about a given concept, which takes place in the social environment of the classroom and also individually, which is derived from the teaching process. The *conceptual profile* is inspired by Bachelard's (1978) idea of epistemological profile, which defends the idea that concepts are in full development, linked to certain philosophical views derived from their stage of maturity; a single philosophical view explains only a fraction of the concept.

The *conceptual profile* shows linearity with ideas from the epistemological profile, such as the division of characteristics between different zones of the profile, each successive zone containing more comprehensive explanatory power than the previous ones. With this, learning is understood as migrating to new zones in a conceptual profile, which do not have to abandon ideas present in other zones, but one must be aware of the interaction between the different conceptual zones and in which context each zone can be used.

The analysis of the conceptual profile guides the perception of the characteristics of a conceptual zone as epistemological and ontological obstacles that hinder the construction of more advanced zones.

In the school environment, although each student has different profiles, their characteristics within the same culture are the same for each concept. In Mortimer's (1996) work, the history of scientific ideas can be used to identify scientific and pre-scientific zones.

When analyzing ideas of heat, Amaral and Mortimer (2001) propose a conceptual profile of heat made up of five zones: realist, animist, substantialist, empiricist and rationalist. These zones are discussed below.

In the realist zone, the idea of heat has an exclusive link to sensations without any thought about the phenomenon, so the construction of ideas does not go beyond sensations. In this zone, common sense ideas about heat and temperature appear.

In the animist zone, the idea of heat is understood as a substance that has life or can generate life, exposing the idea that matter has the will to give or receive heat. This thinking is identified in various contexts, although this view was overcome in the 19th century, it is still identified in school contexts.

The substantialist zone, which was identified by some scientists in the mid-19th century, is currently still identified in everyday teaching, where heat is understood as a substance that has the property of penetrating other materials, despite scientific advances this obstacle has not been overcome.

In the empirical zone, the divergence between the ideas of heat and sensations made it possible to develop the concept of specific heat and the differentiation between heat and temperature, which was made possible by the development of the thermometer, which helped to carry out experiments in which heat could be measured.

The rationalist zone represents a *"body of notions and no longer just as a primitive element of immediate experience"* (BACHELARD, 1978). In this zone, the concept of heat is understood as the relationship between temperature difference and specific heat.

On teaching occasions, the same student may present two or more ways of expressing their understanding of the concept of heat, depending on the context in which they are placed. This means that organizing ideas from the perspective of the conceptual profile will make it easier to understand that concepts have been understood in different ways during their development. The *conceptual profile* has the potential to help the teacher identify the obstacles to learning concepts that the students present.

So we don't want students to change their way of communicating in society, and instead of saying *"I'm hot"* say *"Wow! How hot the environment is and it's transferring energy in the form of heat to my body more than I'm used to"*. But we want to make them aware and enable students to understand that there is another way of thinking about this concept (scientific thinking) and that students can use one way or the other depending on the demands of the situation.

As a result, there is a need for more flexible teaching that meets the social needs of students so that they realize that chemistry makes sense in their daily lives. What is needed is a construction of knowledge that promotes conscious decision-making and enables students to act as critical citizens and modifiers of their *"status quo"*.

1.4 Teaching and learning situations and teaching moments

This work used teaching and learning situations that make up the pedagogical moment of *initial problematization*, which should not be confused with the dynamics of *"study situations"* (with representative works developed by MALDANER, 2007; AUTH et al., 2004; MALDANER; ZANON, 2004) that are based on Vygotsky's ideas, the technique of study situations is very similar to the dynamics of pedagogical moments.

The main difference between the two dynamics is that in the *study situation* the themes addressed are not evident, while in Freire's ideas thematic investigation is the dynamic for acquiring the themes that will be developed. Gehlen, Maldaner and Delizoicov (2012) published a paper that explores the characteristics of each technique and the complementarity of *teaching moments* and *study situations* in the classroom.

The teaching and learning situations in this work are didactic-pedagogical resources that seek to develop in more detail the pedagogical moment of *initial problematization.*

The teaching situations are aimed at the teacher and focus on guiding the teacher's work . The learning situations are aimed at the students and are intended to create situations that bring the student into contact with the subject to be covered. This makes it possible to investigate students' previous conceptions of the subject. The learning situations developed in this work contain situations that can be experienced by students in their daily lives, but also include elements of the history of science.

These proposals are based on work grounded in Freirean ideas, among which the works of Delizoicov (1982, 1991, 2008); Silva (2004); Delizoicov; Angotti; Pernambuco (2002); Pontuschka (1993); Pernambuco, (1994) and Angotti (1982) are worth highlighting.

Based on the proposition presented in the book "Pedagogy of the Oppressed" (FREIRE, 2005), Delizoicov (1991, 2008) proposes the systematization of Freire's thematic approach.

From Freire's perspective, education should be seen as an incessant, restless and, above all, permanent process of seeking knowledge. Freire sees two forms of education. The first, called "Banking", is characterized by the uncritical and apolitical transmission of knowledge and assumes that knowledge is a gift from those who think they know to those who know nothing. The second form is "problematizing" education, in which the teacher must arouse in the students a critical spirit, curiosity and non-acceptance of knowledge that is simply transferred (Freire, 2005). Educators have *"as one of their fundamental tasks to work on methodical rigor with students who must approach cognizable objects"* (Freire, 2006). Learning takes place through the formulation and reformulation of knowledge by students alongside teachers, who are also subjects in the process.

However, transposing Freire's ideas to formal education is problematic, since Freire's theory was basically developed from informal education. In order to facilitate the transposition of Freire's ideas into everyday science classroom activity, Delizoicov structured three pedagogical moments: (I) initial problematization, (II) organization of knowledge and (III) application of knowledge.

1.4.1 Initial problematization

The pedagogical moment of Initial Problematization, according to Delizoicov (1991, 2008) and Delizoicov, Angotti and Pernambuco (2002), consists of presenting real situations that students know and experience, with the aim of encouraging students to be critical when faced with the interpretations of the situations proposed for discussion and making them accept the need to construct new knowledge, with which they can interpret the situation more appropriately. In this way, students are challenged to present their understanding of certain significant situations which are manifestations of local contradictions and which form part of their experiences (FREIRE, 2005).

During the initial problematization, the teacher's task is only to diagnose what the students know and think about a given situation. The student should be free to take part in the discussion and the teacher should not provide ready-made explanations, but should provide the opportunity to seek out questions and interpretations made by the students. At this initial stage, discussion in small groups is recommended, to be shared later in the large group (DELIZOICOV, 2001).

1.4.2 Knowledge organization

In the ideas of Delizoicov (1991, 2008) and Delizoicov, Angotti and Pernambuco (2002), the systematization in the form of a study of the knowledge involved in the theme of the initial problematization is necessary in the second stage of the pedagogical moments. The scientific knowledge necessary for a better understanding of the topic and significant situations is then studied. At this point, it is necessary to show that scientific knowledge is the starting point for new discussions.

The scientific understanding of situations problematized by students is necessary. With this in mind, the teacher takes on the role of developing various activities such as the use of science communication texts, as suggested by Alvetti and Delizoicov (1998). However, this alone is not enough. There is also a need to develop other activities which, according to Ribeiro and Martins (2007), can be related to written production involving narrative. Giordan (2006) highlights the need to use information and communication technologies, which are currently referred to as digital information and communication technologies, due to the ever-increasing presence of these technologies (among which cell phones and notebooks are worth mentioning) in our daily lives.

1.4.3 Application of knowledge

This moment is designed to use the knowledge that the student has been building to analyze and interpret the situations proposed in the initial problematization and others that can be explained and understood by the same body of knowledge.

At this point, the teacher's role is to develop various practices that enable students to use the scientific knowledge explored in the organization of knowledge, with the aim of preparing students to constantly work with scientific conceptualization in situations that are part of their daily lives. At this point, the aim is to identify and use the scientific conceptualization involved, and to explore the explanatory and awareness-raising potential of scientific ideas. From this point on, the student is in a position to scientifically understand the situations addressed in the initial problematization, which is why, in this third moment, they return to the initial situations, which are now understood from the point of view of science.

1.5 Information and Communication Technologies in Education

Accepting that we are in a scenario where social life is highly technological and assuming technology as a necessity in our daily lives, it is necessary to use it in educational practices, seeking greater student interaction with the topic being addressed, taking into account that technological devices such as cell phones and computers can be used as teaching tools.

Today's students are so dependent on technological devices that if they are not used in the classroom, the lesson is worthless. As a result, there is a need for practices that are planned and oriented towards the use of these new technologies that are appearing on a daily basis. These technologies make it possible to transform the way we interact, changing behaviors and relationships and breaking down paradigms of relationships between individuals.

Among the various sectors of society, education is one of the areas that is being affected by this technological wave (FERREIRA, 1998). As a result, we are also facing a time of review of school education, its role and its scope.

The current panorama is one of a revolution in information and communication based on new technologies, also known as ICTs (Information and Communication Technologies), which are updated all the time and which minimize cultural and geographical obstacles, guiding us towards new processes of production, living, thinking, acting and interacting. Understanding education as the basis for the formation of citizens, the need arises to prepare its professionals to master the educational potential that technology offers and to make it available in favor of a pedagogical development that directs its practices towards the construction of the autonomy of the students and the full formation of the exercise of citizenship (SOLANGE, 2011).

The National High School Curriculum Parameters (PCNEM) guide the use of computers in education as a tool for new learning strategies, capable of making a significant contribution to the process of building knowledge in various areas (BRASIL, 1999).

One of the main contributions of ICTs is to combat educational problems, as they are capable of magically transforming the traditional teaching process into real show classes and, in this way, combating students' lack of interest in the content. Thus, the quality of education would be achieved through the use of technological resources that would make teaching more attractive (MOREIRA and KRAMER, 2007).

This discourse by academics in the field of Education and Science Education has generated different reactions. In science education, authors generally defend the advantages of ICT (for example, GIORDAN, 2005; TONIATO et al., 2006; SERRA and ARROIO, 2007) as new teaching resources that update and facilitate teaching, or that motivate or contribute to the science learning process. In these works, computer resources are used for students to research topics, prepare graphs, spreadsheets, obtain data in experiments, visualize images and simulations with the aim of motivating or "modernizing" the lesson.

The use of computers in education, especially in chemistry, has contributed considerably to the teaching and learning process. As an experimental science, it has an observational character. Most of the theories used to explain chemical reactions require models, such as the quantum model of the atom, molecular orbitals, among others. The help of computer simulations and animations, interactive programs, among others, can help students understand the model better, overcoming some of these barriers.

In the field of Education, on the other hand, we find authors who problematize the insertion of ICT in Education (for example, PRETTO, 2001; BARRETO, 2009), drawing attention to the necessary (re)contextualization of these artifacts, since they were created at a given historical moment and for specific purposes.

ICT should not be used as a machine for teaching or learning, but as a pedagogical tool to create an interactive environment that allows students, when faced with a problem situation, to investigate, raise hypotheses, test them and refine their initial ideas, thus building their own knowledge.

CHAPTER II: METHODOLOGY

2.1 The research profile

This investigation is characterized as pedagogical research (LANKSHEAR and KNOBEL, 2008) which, as such, arises from the intention of being an antagonist to the tendency to study terms referring to the daily life of the classroom. The development of this type of study requires the involvement of players in the teaching and learning process, knowing who the researcher is, in this case, the teacher whose object of study should be the classroom.

The research was mediated by the qualitative research methodology, which is not concerned with numerical representativeness, but rather with deepening the understanding of a social group, an organization, etc. Researchers who adopt the qualitative approach oppose the assumption that there is a single research model for all sciences, since the social sciences have their own specificity, which presupposes their own methodology. Thus, qualitative researchers reject the positivist model applied to the study of social life, since the researcher cannot make judgments or allow their prejudices and beliefs to contaminate the research (GOLDENBERG, 1997, p. 34).

These ideas are also shared directly and indirectly by Minayo, (2003 and 2007); Bauer, (2002); Trivinos, (1987); Caregnato, (2006); Flick, (2004); Lüdke, (1986); Moreira, (2002); Demo, (1991 and 1995); Fidel, (1992), Gil, (1991); Alves-Mazzotti and Gewandsznajder, (1999).

2.2 Methodological paths

This research was carried out at the Imaculada Conceiçao State School of Primary and Secondary Education, in the Portal do Poço neighborhood, in the city of Cabedelo, in the state of Paraiba, with 40 students from the 2nd and 3rd years of regular secondary education. In these classes, the pedagogical moment of Initial Problematization was applied, based on the BBC documentary entitled "Absolute Zero: the Conquest of Cold (Episode 1)", available at https://www.youtube.com/watch?v=jtLklcHVHic, whose main ideas are set out in Appendix 4. All the classroom interventions were recorded on audio and video to be transcribed later.

The documentary ran for 29 minutes and 13 seconds. Using this excerpt from the documentary, two learning situations (LS) were produced for the students (shown in Appendices 1 and 2), which contained texts related to the concepts of heat and temperature, including historical, contextual, contemporary, cultural, political and social aspects. In addition, the SA contained investigative questions whose main objective was to explore the students' conceptions of some historical, social, economic, political and contemporary issues in the evolution of the concepts of heat and temperature.

Complementing this approach, a teaching situation (SE) was produced for the teacher, available in

Appendix 3, containing guidelines on how the teacher can organize the classroom and how to work with the class, some complementary concepts of heat and temperature that are not normally covered in much of the literature intended for teaching, some possible answers to the questions in the learning situations, a pre-test to be applied when the teacher sees fit, as well as some competences and skills from the reference matrix for Natural Sciences and their Technologies from the 2013 Enem.

The learning situations were applied in six 45-minute lessons. In the first lesson, the students were asked to form groups of four and were instructed to read the texts and discuss them in order to prepare the answers to the questions, which should contain the ideas of the group and not just one or other component. The following day, two 45-minute lessons were used to administer a pre-test (set out in item 4 of the teaching situation), with objective multiple-choice questions, covering situations that can be experienced in the students' daily lives.

The data obtained by applying the SA and the pre-test contained in the SE were analyzed from the point of view of the *"conceptual profiles"* (already discussed in section 1.3 of chapter 1), as a way of guiding students through the need to obtain new knowledge and new ways of seeing and understanding the world in which they live. In the analysis, the intention was simply to identify conceptual cognitive barriers that should be taken into account when explaining the content in class, with a view to overcoming these barriers.

Conceptual cognitive barriers were identified by comparing the ideas expressed in the students' answers with current scientific ideas.

CHAPTER III: RESULTS AND DISCUSSION

3.1 . Learning situations produced

The two learning situations (LS) were created to carry out the initial problematizing pedagogical moment, with the aim of probing the students' previous conceptions and world knowledge about the topic to be taught (concepts of heat and temperature). This probing should be done indirectly and should not directly address the subject, as in the ideas of Delizoicov (1991, 2008) and Delizoicov, Angotti and Pernambuco (2002).

The SA of this work was created in text form with the main ideas based on the BBC documentary entitled "Absolute Zero: the Conquest of Cold - Part 1" up to the time of 29 minutes and 13 seconds. In the first SA (Appendix 1), the text takes a look at the mysticism surrounding cold and how some phenomena were understood in ancient times. It also looks at a period of great drops in temperature that took place in England in the 17th century and how this affected people's social context at the time, talks a little about Drebbel's experiment that possibly cooled the air in Westminster's Great Hall and the contribution of volcanic eruptions to the drops in temperature.

In the second SA (Appendix 2), the text discusses Boyle's curiosity about cold and some experiments he carried out to investigate the nature of cold, as well as the evolution of thermometers and their use throughout history, Lavoisier's caloric theory and the two views of heat, (I) Substantialist and (II) Mechanical. The texts deal with historical, social, cultural and geographical conceptions of the construction of ideas about the concepts of heat and temperature. Both guided an assignment in which the students had to form groups of 4, read the text and the statements, debate and formulate the answers to the statements, which included some elements of contextualization.

3.2 Application of SA

During the application of the SA, some facts emerged that are worth highlighting, since it is believed that these facts influenced some of the answers that will be highlighted and commented on in the next section.

When they arrived at the classroom to apply the SA, the SE guidelines were followed (Appendix 4), it was explained to the students that they should get together in groups of four to answer the questions based on two texts that would be given to the group, and that they shouldn't put their names on the SA sheets, because we weren't interested in knowing the idea or understanding of one student or another, but rather the ideas and understanding of the group of students who took part in the research, since the lessons would be given to all the students and not to one student in particular.

The students were also instructed not to search for the answers in any source other than the SA texts;

the answers should be constructed from a debate between all the group members to express the group's idea and not just one of the group members.

When I presented the activity to the students, they asked how many points the activity would count towards their bimonthly grade. When the students were told that it wouldn't count towards their grade, because the purpose of the activity was to check their understanding of the topic, there was a huge outcry. The students harshly questioned the way the activity was carried out because it wasn't worth points towards the bimonthly grade, and then said that the teacher's duty would be to explain the subject on the board, do and answer examples on the board and choose five questions and change the numbers to take the test, using expressions such as "*this is the kind of teacher the government sends us, where are we going to get this way?* ", or even, "*Which college are you buying your diplomas from? This is absurd*".

After questioning the students, they were told that those who didn't want to do the activities shouldn't do them, as they wouldn't be harmed. However, they all agreed to do the activities.

Despite the instructions given to the students, none of the groups followed them correctly. It was noted that most of the students were researching the answers on the internet. In some groups, the members of the group divided up the answers to the questionnaire, with each member being responsible for answering three or four questions. Some groups split into two sub-groups, with each sub-group responsible for answering the SA questions. In other groups, only one of the students was answering the questions.

It is therefore believed that most of the answers do not express the students' real understanding of the subject, but rather some answers that originate from internet searches or answers to statements expressing the ideas and understanding of just one of the group members and not an idea built on a debate between all the members.

3.3 Students' answers to the SA

This section contains literary transcriptions of the students' SA answers. However, the discussion in this section will not take into account the spelling mistakes and errors of verbal and nominal agreement and regency made by the students, only the ideas they tried to express in their answers to the SA questions.

3.3.1 Students' answers to the first SA

3.3.1.1 First statement: everyday experience of the cold

The following are the ideas contained in the students' responses to the first statement of the first SA, in which the students are asked to answer, based on their everyday experiences, what cold is and what its effects are.

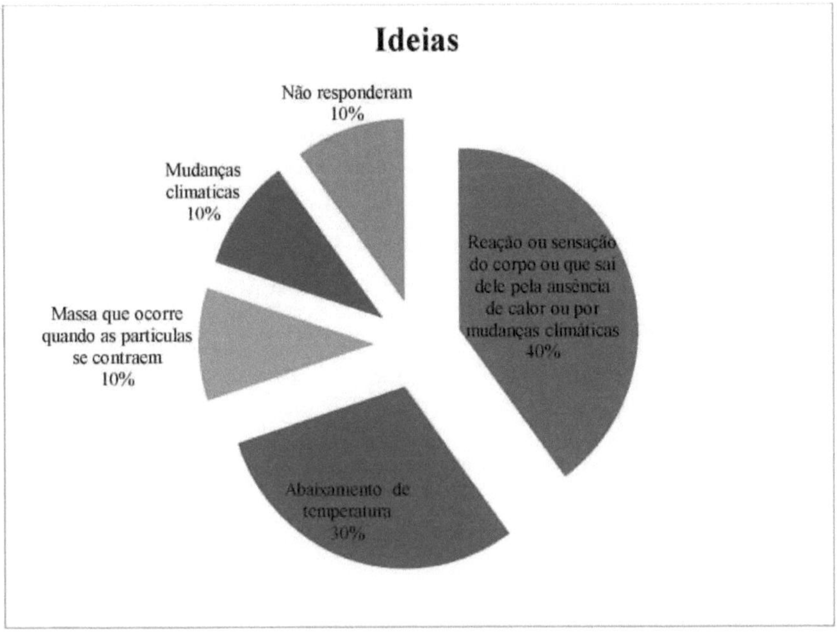

Figure 4 - Students' answers to the first statement of the first SA.

None of the answers expressed a scientific idea about the cold; the answers about what the cold is were statements of what they believed to be their everyday experiences or ideas that had been passed down from their older relatives and members of their community. These answers relate the cold to climate change and a drop in temperature.

The students' ideas of the effects of the cold are closely linked to their experiences. They say that the effects of the cold are the flu, chills on the skin, illness, rain, floods, Ieptospirosis, fatigue, sleep, among others.

These ideas put forward by the students come from common sense, those ideas passed down from father to son, ideas shared in their geographical and social environment, and they also demonstrate colloquial language.

We can see these ideas in the transcript of the answer below:

"Cold is a sensation that comes out of our body when we come into contact with wind, water, ice or anything like that. Its effects are chills, the body freezes, the lips turn purple and the body shakes and the cold itself can lead to death. "

The transcripts below show answers that were probably taken from an internet search, where heat is clearly treated as a substance (cold would be the "*lack of heat in a body*"), but the ideas are at odds with current scientific concepts (bodies don't have heat). However, in colloquial language, we can come across these distorted concepts in everyday life, in social circles, even by people in academia.

"Whether it was a substance, a process, or some special state of existence is not known for certain, a real force is not identified but questioned, and the effect is an extremely cold air that can cause freezing effects."

"Cold is the sensation produced by the lack of heat in a body or material, caused by low atmospheric temperature or by artificial means through refrigeration. The effects of cold can cause burns or hypothermia, which sharply lowers body temperature, and at a less severe level can cause fatigue, limpness, shivering, cramps and pain."

In the first transcript below, the students understand cold as a mass, which clearly expresses an idea of cold as a substance. In the second transcript below, students understand cold as the change in climate when the weather turns rainy or foggy.

Although the students' answers are inadequate from a scientific point of view, the second transcript below relates cold to climates with low temperatures.

"The cold for me is a cold mass that occurs when the particles contract and become cold and when it gets too cold it leads to catching a cold"

"Cold is a climate change that occurs during rain, cloudy weather or the absence of heat. Asthma, a cold, a stuffy nose and a lack of heat are just some of the effects."

3.3.1.2 Second statement: Drebbel's device

Analyzing the students' answers to statement 2 of the first SA, which asks students to draw a diagram of a device used by Drebbel to cool the air, a device proposed by Dr. Szydlo, where the way to use the device is described in the text of the first SA.

None of the groups expressed a satisfactory idea in the form of a drawing. Some groups expressed unsatisfactory ideas of Dr. Szydlo's proposal by placing the ice and salt in separate containers, others by placing a container with hot air and cold air on the other side, some drew only a fan and another group even drew a lighthouse, while another group drew an air conditioner. Figure 4 shows the closest idea to the machine used by Drebbel.

It can be seen that the text in figure 4 presents an idea that converges with those in the SA text, while the drawing is placed in a way that diverges from the ideas in the SA text.

Statement 2

Based on the information contained in the text, check out the device proposed by Dr. Szydlo Siipostaincnte and used by Drebbel to cool the air.

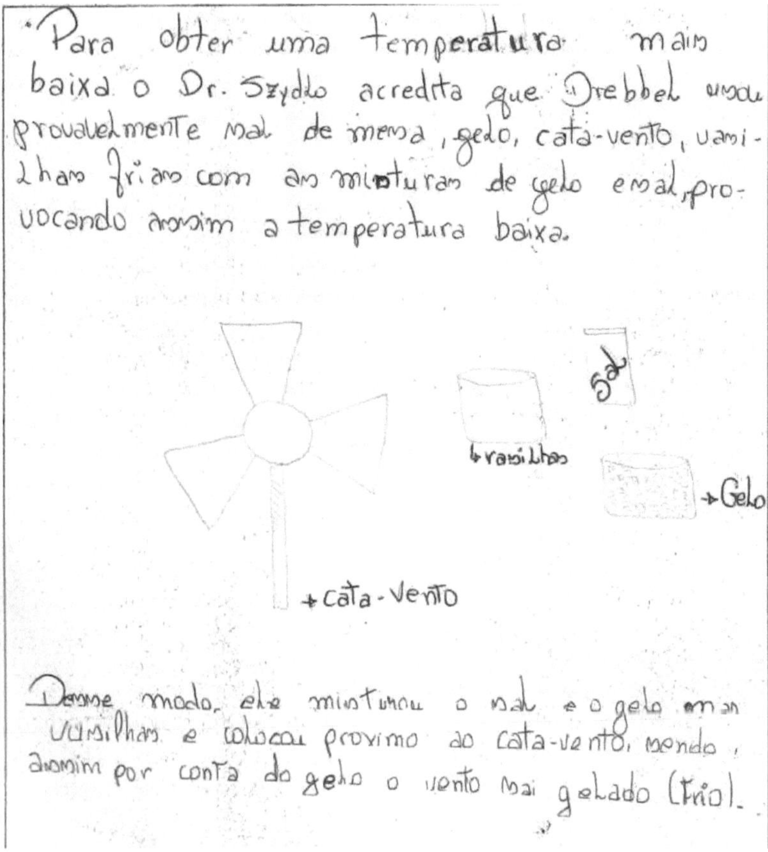

Figure 5 - Students' response to the second statement of the first SA.

Figure 5 shows a drawing that is possibly a schematic of air conditioning equipment taken from the internet and the group of students merged this schematic with the ideas contained in the SA text to give a vision of the Drebbel machine suggested by Dr. Szydlo.

Statement 2

Based on the information contained in the text, draw a diagram of the device proposed by Dr. Szydlo and supposedly used by Drebbel to resinate air.

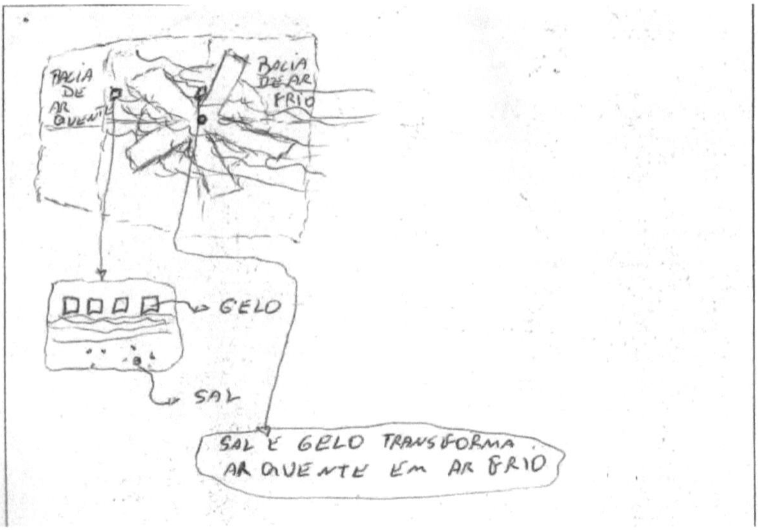

Figure 6 - Students' answers to the second statement of the first SA.

In figure 6, the students literally draw an air-conditioning machine of the earliest type. The name *"air-conditioning" is* also explicit, which shows that the students who drew this figure were unable to understand the part of the SA text that commented on the device created by Dr. Szydlo, which Drebbel supposedly used to accomplish his great feat.

In figure 7, the students draw a container with hot air and another with cold air and believe that when they mix the ice and salt there is a kind of transformation from hot air to cold air.

Statement 2

Based on the information contained in the text, sketch the device proposed by Dr. Szydlo. supposedly used by Drebbel to cool the air.

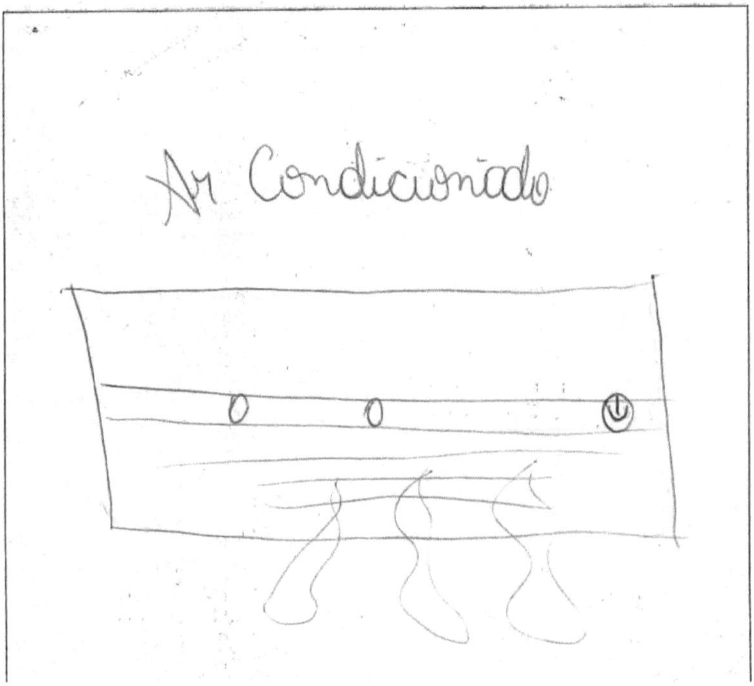

Figure 7 - Students' answers to the second statement of the first SA.

Statement 2

Based on the information contained in the text, draw a diagram of the device proposed by. Dr,. Szydlo supposedly used by Drebbel to resinate air.

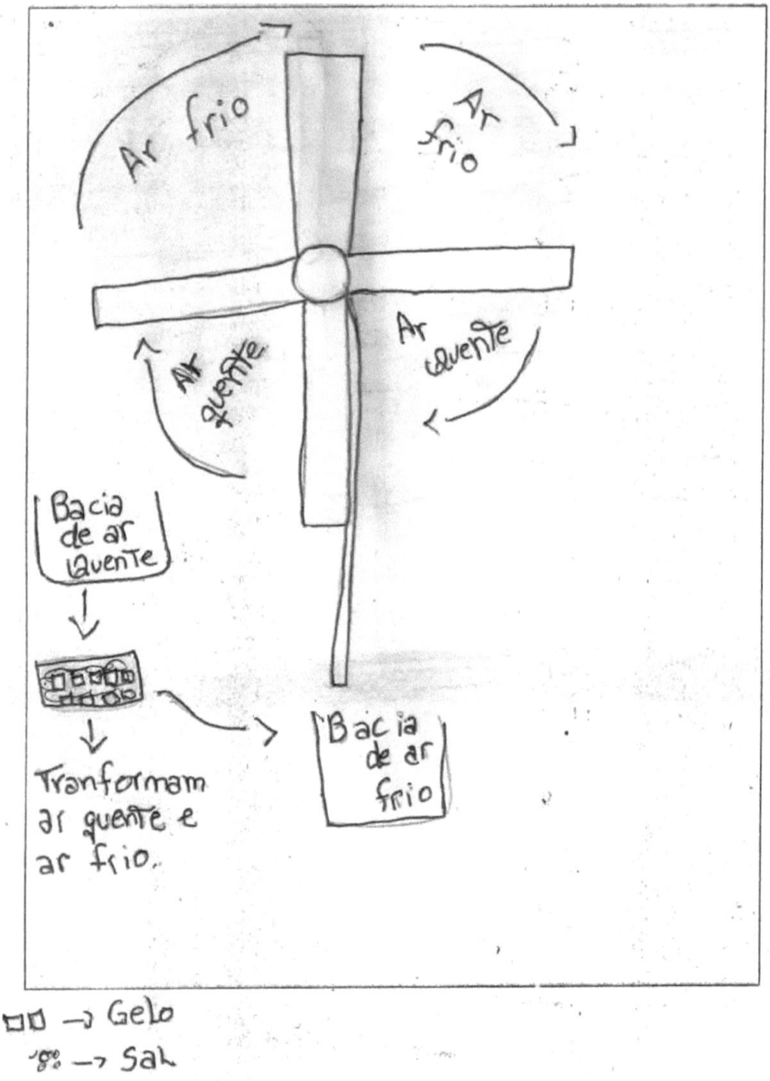

3.3.1.3 Third statement: origin of phenomena

Analyzing the students' answers, the following ideas were identified in relation to statement 3 of the first SA, in which the students had to explain their understanding, arguing whether phenomena that were previously understood as divine or diabolical phenomena were still understood in this way today.

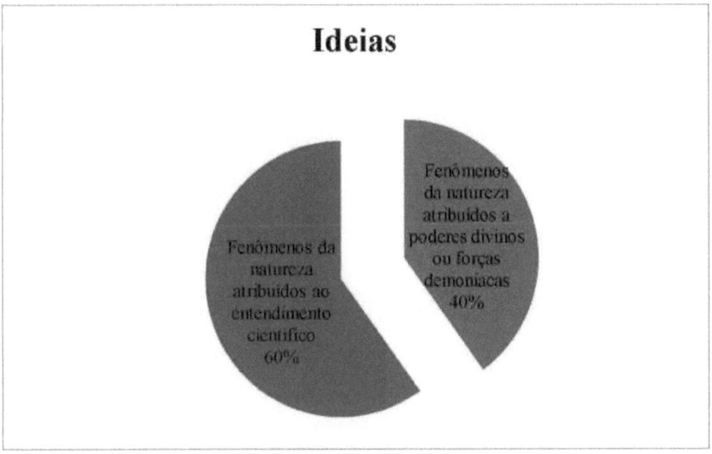

Figure 9 - Students' answers to the third statement of the first SA.

Whenever the subject involves beliefs and divinities, the ideas expressed are the most varied. In this section it is important to emphasize that this work does not mention, idolize or defend any deity or religion, nor is it intended to attack any type of belief, we are only interested in knowing how people understand the phenomena that occur in their time, and if this understanding of phenomena is still like that of people who lived centuries ago or if it has undergone some kind of reformulation.

As examples of these phenomena, the students cited tsunamis, lightning, thunder, earthquakes, heat, cold, among others.

Looking at figure 9, we can see that 60% of the students have reformulated the way they understand some of the phenomena of their time when compared to the understanding of people who lived in previous eras. It is believed that this reformulation of thinking may have been the result of the advance and dissemination of science and technology in recent decades, which are becoming increasingly present in people's lives.

Examples of these ideas are the transcriptions of the answers below. In the second transcript below, cold is expressed as "a mass", demonstrating a substantialist idea of cold.

"Today they are seen as facts easily explained by scientists and attributed to global warming, fires, eruptions, deforestation, air currents and various other factors."

"Science today can already prove the true cause of some phenomena. For example, thunder is not a punishment from God, it's just a mass of hot air colliding with a mass of cold air, thus creating thunder and lightning".

Figure 9 shows that 40% of the students have not reformulated their ideas about the attribution of the phenomena that occur in their time, when compared to the ideas of people who lived in past times, and continue to attribute the phenomena to divine or demonic forces without any kind of human participation or influence. Examples of these ideas can be found in the transcripts of the answers below.

"Depending on the phenomenon, if it's a good thing, people say it was divine power, if it's a bad phenomenon like hauntings or tsunamis, or even an earthquake or whirlwind, they think it's demonic. But I think these things exist because of human beings who want to know more than God, who want to work miracles, underestimating God's power."

"Nowadays, natural phenomena are seen as the end of times, the wrath of God."

3.3.1.4 Fourth statement: the cold and history

Analyzing the students' answers, the following ideas were identified in relation to statement 4 of the first SA, in which the students had to reflect and argue whether the cold could change the course of history.

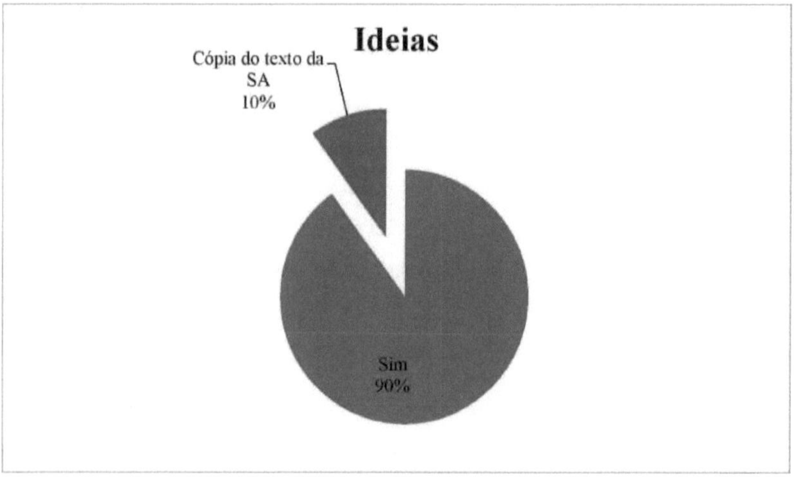

Figure 10 - Students' response to the fourth statement of the first SA.

Most of the answers were satisfactory and agreed with the ideas in the SA text and with some current scientific ideas, expressing ideas such as: the cold could change the course of history and they highlighted natural and social phenomena that would be caused by the cold, stating that many living beings would die, technology would be renewed, crops would freeze, food would become scarce and more expensive, which could lead to attacks and riots in warehouses and markets.

A group equivalent to 10% of the students cited the movie "The Day After Tomorrow", which portrays global warming causing extreme climate change, devastation and an ice age on the planet.

These ideas contained in the answers are transcribed below:

"Yes, the movie 'The Day After Tomorrow' shows what would happen if the world became so cold that it would suffer chaos. Farmers would suffer because they wouldn't be able to plant or harvest anything because of the cold. People might

even stop going out so much because every time they went out they would have to put on lots of warm clothes'.

"Yes, because it would drastically change the economy, the climate, the way of life and these factors would change the course of history.

"Yes, because with every study, new discoveries are made, and with that, other inventions, which means that both technology and history are renewed."

The answer transcribed below, while agreeing that cold can change the course of history, brings a totally different idea from the ideas contained in SA's text, going so far as to absurdly state that if it weren't for the cold seasons the volcano wouldn't evaporate.

"Yes, because the cold spells at that time had an influence on the volcanic increase, because if there hadn't been these cold spells the volcano wouldn't have evaporated."

The answer transcribed below is an excerpt from the SA text, so this answer does not express these students' ideas about the statement.

"Although there is no consensus, some climatologists believe that it was soil agitation"

3.3.1.5 Fifth statement: volcanic eruptions

Analyzing the students' answers, the following ideas were identified in relation to statement 5 of the first SA, in which the students had to argue, based on the information contained in the text, how volcanic eruptions influenced the drop in temperature.

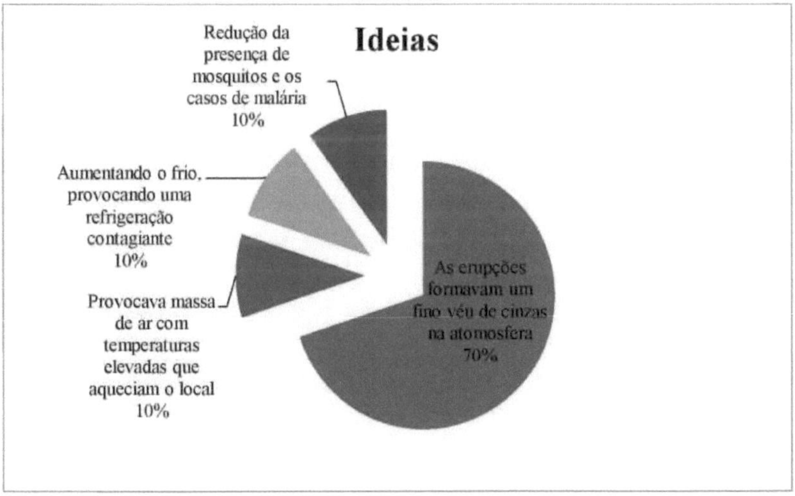

Figure 11 - Students' response to the fifth statement of the first SA.

In this statement, we expected a high number of answers with ideas that converged with those in the SA texts, but only 2 groups responded accordingly, while other groups expressed answers with ideas contrary to those contained in the SA text and statement.

In the transcript below, the students answered with well-developed information, not contained in the text (it is believed that these answers came from an internet search):

"The cooling is explained in a simple way: the volcano releases large quantities of volcanic ash and sulphur dioxide, which are transported to the stratosphere, closest to the surface. There, physical and chemical phenomena create a thin layer of whitish particles which, for months or years, surround the earth and reflect part of the sun's rays, preventing radiation from reaching the ground."

In the two transcriptions below, although the students expressed an idea that agreed with that of the SA text, they failed to realize that the correct answer is in the antepenultimate paragraph of the SA text, and added part of the previous paragraph.

"Although there is no consensus, some climatologists believe that it was solar agitation. The increase in volcanic activity, whose eruptions covered the upper layers of the atmosphere with a thin veil of ash, may also have had an influence."

"Some climatologists believe it was solar agitation. The increase in volcanic activity, whose eruptions covered the upper layers of the atmosphere with a thin veil of ash, may also have had an influence."

The transcript of the answer below shows an answer that satisfactorily summarizes the ideas of the SA text.

"Because the ashes made a kind of cloak and prevented the sunlight from getting through."

The transcript of the answer below contains some excerpts from the SA text, but they are loose excerpts that don't make sense to the answer.

"They influenced the reduction in the presence of shrews and cases of malaria, although it doesn't seem like much, it was enough to transform the lives of millions of people."

In the two transcriptions below, there are different ideas from those mentioned in the statement: in the first transcription, instead of the group arguing about the influence of volcanic eruptions on lowering the temperature, they argued about how volcanic eruptions influenced the rise in temperature. In the second transcript, the group simply argues that there is excessive cold that spreads quickly.

"The subsidence is due to the lava spewing out of the volcano at very high temperatures, and the volcano also causes a mass of air at very high temperatures which heats the place up completely."

"Because the exertion of the cold is increasing and it is gaining a lot of strength and causing a contagious chill that is spreading rapidly."

3.3.1.6 Sixth statement: cold and economy

Analyzing the students' answers, the following ideas were identified in relation to statement 6 of the first SA, in which the students had to discuss how high temperatures can affect a nation's economy.

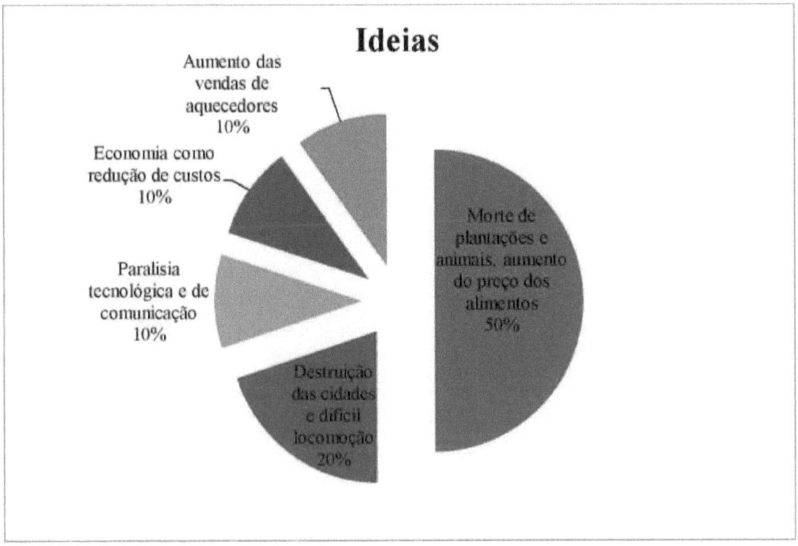

Figure 12 - Students' answers to the sixth statement of the first SA.

Most of the students' answers showed ideas that are in line with current scientific ideas, which in this case are very similar to common sense ideas, as they are situations often not experienced by the students, but which are frequently circulated on television and electronic media.

In the transcript below, the group of students expresses an adverse idea of the meaning of economy contained in the text and in the statement, treating the economy of a nation as merely a reduction in spending and not as *"the set of activities carried out by men aimed at the production, distribution and consumption of goods and services necessary for survival and quality of life"* (FEAUSP, 2008).

"Sometimes they do and sometimes they don't because they don't always happen to a nation's economy."

In the transcript of the answer below, the group of students say that crops and animals would die if it rained constantly. However, this is one of the situations necessary for crops to grow and, consequently, for people and animals to obtain food. It is believed that the students confused the term constant rainfall with flooding.

"High temperatures cause a major crisis in a nation's economy because crops and animals die due to the constant rain. As a result, the population is unable to sustain itself due to the high cost of food, which has to be exported from elsewhere."

In the transcript of the answer below, the students put forward an argument that could be considerable, explaining that high temperatures would increase the sale of heaters. However, when you look at the bigger picture, this situation would be insignificant compared to what would happen with the rest of a nation's economic factors.

"The high temperatures could affect a large part of a nation's economy, but it could also help because today, with the cold weather on the rise, sales of heaters could increase."

In the transcript of the answer below, the concept of heat still appears with the substantialist conception of centuries past, taking up ideas similar to those of Lavoisier's caloric and other scientists of that time. It is also clear that the students in this group associate summer only with hot seasons and not as a season of the year that refers to a region of the Earth that is more exposed to the sun at a given time.

"Because if it gets too cold, crops in general won't thrive, and there are also many people who depend on the summer on the beaches with bars that only work if it's hot.

In the transcript of the answer below, the group of students puts forward an interesting argument. While the other groups report only that the cold would affect the economy of a region, this group argues that this will happen in a society that is not used to or prepared for the cold.

"With the arrival of the cold the economy would be affected, in a nation that is not used to a cold that has never happened in their place of residence, it would affect the prices of food and so on, everything would become more expensive, because it would increase the prices to survive in this climate, that's why it would affect the economy of a country."

3.3.1.7 Seventh statement: technology against the cold

Analyzing the students' answers, the following ideas were identified in relation to statement 7 of the first SA, in which the students had to justify their answers to the question of whether current technology could prevent or minimize the effects of a new ice age.

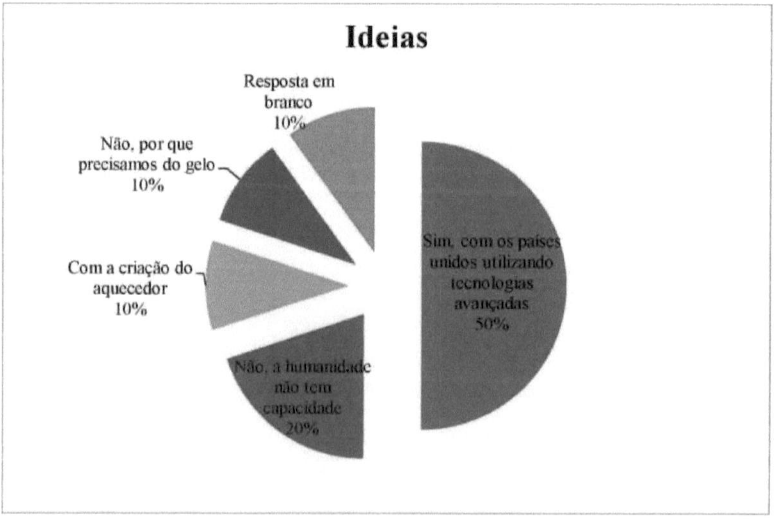

Figure 13 - Students' answers to the seventh statement of the first SA.

Half of the students believe that current technology could prevent or minimize the effects of a new ice age. However, most of these answers were poorly formulated, some answers only said yes and repeated a part of the question in the answer as a means of justifying them.

In the two transcripts of the answers below, the students present a well-established argument by stating that with current technologies there would be no way to avoid the effects of a new ice age, but there would be a way to prevent and minimize the effects of a new ice age.

"Due to the countless technologies we have and the technological advances we have today, I would say that if the powers of our world and our countries were to unite, there would be a way not to avoid, but to mitigate and prevent such a catastrophe."

"Yes, because with today's technologies it would be quicker to avoid or minimize the effects of a new ice age."

In the two transcriptions of the answers below, the students expressed ideas that show an interpretation of the statement as if current technology would be able to prevent a new ice age, although they were expecting an argument that revolved around minimizing the effects of a new ice age.

"No, because if they haven't come up with something that can stop some natural phenomena, how can they stop the ice age?"

"Not because humanity doesn't yet have the capacity to prevent or minimize a phenomenon of this size."

In the transcripts of the answers below, the students say that Drebbel created the first air conditioning and with today's technology someone could do the opposite of what he did. These ideas do not appear in the SA text; in the SA text it says that if Drebbel had registered his great number, he could have gone down in history as the inventor of air conditioning, and a piece of equipment using an idea contrary to Drebbel's has already been invented and we know it as a heater.

"Yes, because the way Drebbel invented the world's first air conditioner, someone else could have the same idea, only the other way around."

"Yes, because as technology evolves, machines can be built to contain the cold and even prevent it for a while."

In the transcript of the answer below, the students express the idea that ice can only be formed naturally, disregarding their knowledge of the refrigerator, which is now a very common piece of equipment in homes.

"No, because most of the time we need ice, so we can't avoid it or minimize it"

3.3.1.8 Eighth statement: the new ice age

Analyzing the students' answers, the following ideas were identified in relation to statement 8 of the first SA, which presented a text with the title "German scientist says new ice age is coming". The text talks about the cold cycles in the world and, based on some data, some groups of scientists believe that we are heading towards a new ice age. Based on their everyday experiences, the students should present their conclusions about how a new ice age would affect the world, not only economically, but also socially and technologically.

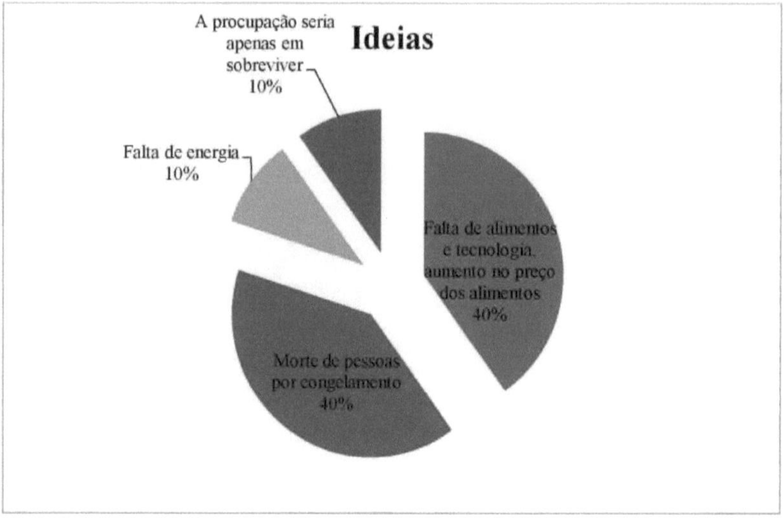

Figure 14 - Students' answers to the eighth statement of the first SA.

The transcript of the answer below is one of 5 answers where the students copied part of the text of statement 8 to answer it.

"Yes, because ice can wreak havoc, displacing entire nations, destroying fragile economies and causing up to a fifth of the world's population to freeze to death."

In the transcript of the answer below, the group of students poses a situation where a new ice age would come so suddenly that people would not have time to take precautions to survive it, preserving some of the economy and technology.

"With the planet freezing, nations would be more concerned with surviving, because with the ice the crops would freeze. I don't think we'd have time to worry about the economy or technology."

The transcript of the answer below contains part of the text of statement 8. In this transcript, the students do not comment on how a new ice age would affect the economy, society and technology. They only answer that technology or nations would not prevent such an event, so these students do not seem to have associated one of the ideas contained in the text, which does not expose the idea that ice ages could be prevented, but only talks about how these cold periods affected a society.

"With reference to the text, the new ice age is fraught with complications. Worse than the first, it will bring enormous destruction, with the fifth part of the population destroying everything in its path. No amount of technology or even the

nation itself will be able to prevent this from happening."

It can be seen that the students have great difficulty in associating social, economic and technological issues with their daily lives, none of the groups argued how a new ice age would affect technology, which shows that although the students live with various types of technology on a daily basis, mastering how they are used, how they work and how they are maintained, they still have difficulty putting forward ideas about how some technologies work and how they would be affected by extreme cold.

3.3.2 Students' answers to the second SA

3.3.2.1. First statement: the thermometer

Analyzing the students' answers, the following ideas were identified regarding the first statement of the second SA, in which the students had to describe in great detail what the thermometer they know looks like.

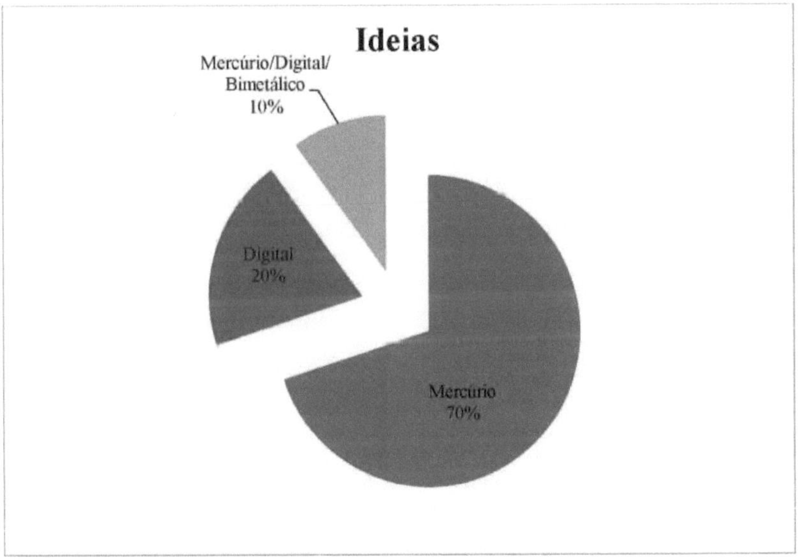

Figure 15 - Students' answers to the first statement of the second SA.

The answers were very diverse, but 70% of the students said they only knew the mercury thermometer, 20% of the students said they only knew the digital thermometer, and 10% of the students said they knew the mercury, digital and bimetallic thermometers.

It was hoped that a greater number of students would say that they were familiar with the digital thermometer, given its current affordability and the ease with which it can be manipulated and the temperature displayed.

The transcriptions of the answers below suggest that, due to the details and the way they are written, this is possibly an answer taken from an internet search. However, in the first transcript, the students mistakenly state that the thermometer controls the temperature, since it is known that the thermometer is only an instrument for measuring the temperature of a given body.

"Premium mercury clinical thermometer of the primastic type, made of very resistant glass material and placed under the armpits, easy to see, with analog measurement given in °C unit, measures fever and controls temperature."

"We've used it but we don't have one. We know one that is called Digital, which is used to check the temperature. This type of thermometer contains a battery or even a battery, it is white, has a nozzle at the end measuring approximately 15cm, and when you are checking the temperature it beeps and tells you how many degrees you are at, regardless of whether it is low or high."

In the transcript of the answer below, the group of students did not say which thermometer they knew, but from the details expressed by the students, it is believed that this is the thermometer known as the mercury thermometer.

"It's made of glass, it's long and has a centimeter metal tip to measure the temperature."

In the transcript of the answer below, the students describe that they know the bimetallic thermometer, a very unusual description since it is not common for people to know this type of thermometer. The most common and popular are mercury and digital thermometers.

"The traditional thermometer has mercury in its composition. And the digital, and the bimetallic thermometer."

3.3.2.2. Second statement: mercury in thermometers

Analyzing the students' answers, the following ideas were identified with regard to the second statement of the second SA, in which the students had to answer why it was important to use mercury in thermometers in the Fahrenheit era.

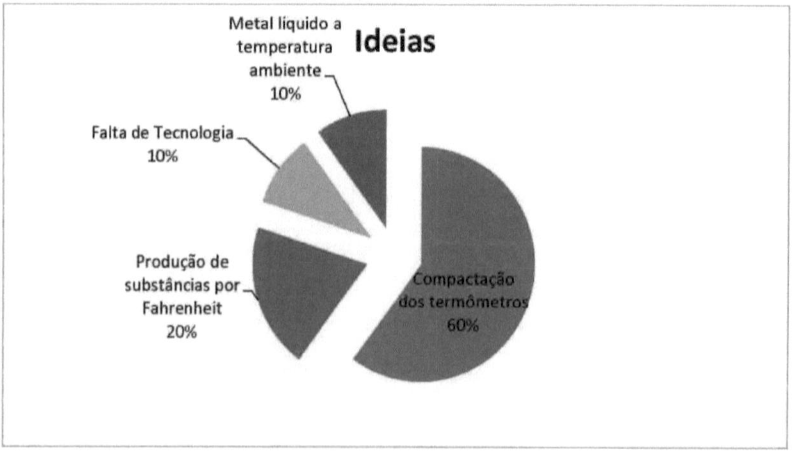

Figure 16 - Students' answers to the second statement of the second SA.

A higher number of satisfactory answers was expected for this statement, as the answer was provided in the SA text. However, 60% of the students answered the statement satisfactorily, with ideas that revolved around the use of mercury to compact thermometers, but they only answered that the thermometers became more compact and for clinical purposes and did not comment on why the thermometers would become more compact with this use.

In the transcript of the answer below, the students state that Fahrenheit created a substance that was accepted for clinical purposes, ideas that do not appear in the SA texts, demonstrating that this group of students was unable to assimilate the information contained in the text.

"The first thermometers used alcohol which is weaker than mercury, the importance of using mercury is because Fahrenheit was able to make a completely different substance than the other scientists and it was accepted for clinical purposes, and mercury is dangerous to man."

In the transcript of the answer below, the students say that mercury was easier to heat, when in fact mercury was used because it is denser and has a higher boiling point than alcohol, so heating it would require more energy, which would make it more difficult to heat than alcohol. The students also claim that mercury is a red liquid, when it is known that mercury has a metallic silver color.

"Because by using mercury, the thermometers were smaller and therefore easier to transport and mercury is a way of making the temperature more specific, because mercury heats up more easily, causing the molecules to agitate so that the red liquid moves and stops at the right temperature."

In the transcript of the answer below, the students only say that in Fahrenheit's time there was no technology. It is believed that the students wanted to express that in Fahrenheit's time no one had the technology to make smaller thermometers and that this feat was achieved by Fahrenheit.

"Because back then there was no technology."

The transcription of the answer below was one of the most consistent with the ideas in the SA text, but the students didn't talk about how the use of mercury influenced the compactness of thermometers.

"The importance of mercury thermometers at the time of Fahrenheit was to be compact and suitable for clinical purposes."

In the transcript of the answer below, it is believed that the students researched these ideas on the internet, and that it only brings the information of the physical state of mercury at room temperature and that mercury has been known since the times of Ancient Greece.

"The importance of attracting mercury in thermometers is because it is a liquid metal at room temperature, known since the time of Ancient Greece."

In the transcript of the answer below, the students state that mercury is various substances that are used for various types of people, ideas that are totally at odds with the ideas contained in the SA text.

"Mercury was various types of substance used for various types of people. For example, they are compacted for some big things on patients."

3.3.2.3. Third statement: thermometer and society

Analyzing the students' answers, the following ideas were identified with regard to the third statement of the second SA, in which the students had to argue about the importance of the thermometer in modern society.

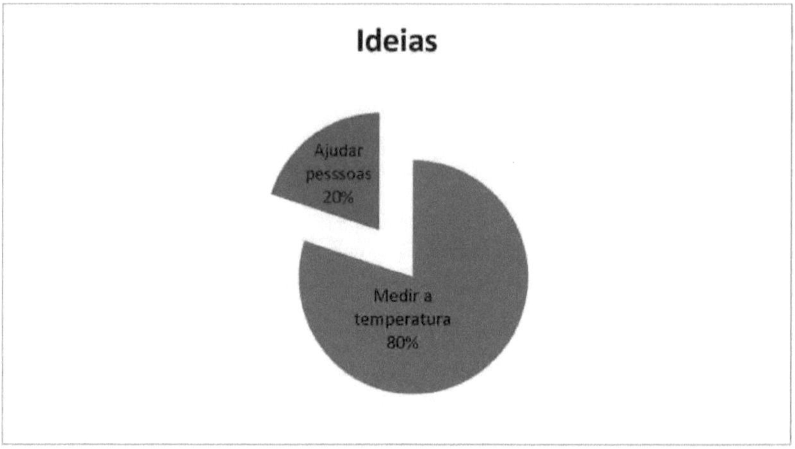

Figure 17 - Students' answers to the third statement of the second SA.

The answers to this question were very varied and limited, but I believe they are in line with the knowledge of high school students. 80% of the students related the importance of the thermometer to temperature measurements. However, there was no mention of the importance of thermometers for society.

In this transcript of the answer *"To control body temperature"*. The students say that the thermometer is used to control the temperature of a body, which is at odds with the ideas contained in the SA text and current scientific ideas.

In the transcript of the answer below, the students only say that the thermometer is important and that it helps many people, but they don't talk about how the thermometer is important.

"The importance is that the thermometer in modern society has a way of helping various people, of course we know that the thermometer is important and very precise in society."

In the transcripts of the answers below, the students express ideas that converge with the ideas in the SA text and with current scientific ideas, by relating the usefulness of the thermometer for measuring temperature.

"The importance, if the thermometer didn't exist we wouldn't know if someone had a low or high body temperature."

"The importance of the thermometer in modern society is that it helps us to know the temperature."

"As well as measuring our body temperature, it's very useful for knowing the temperature of the environment."

In the transcript of the answer below, the students only check the temperature when someone has a fever. There is also the idea that the thermometer assesses how someone is doing, an idea that appears to be mistaken since it is known that it is the doctor or some other health professional who assesses the state of health.

"The importance of the thermometer is to check your temperature when you have a fever, preventing the worst from happening. The thermometer is not like any other object, it is of great importance in modern society because it assesses how you are."

In the transcript of the answer below, the students express a mistaken or poorly formulated idea by stating that the thermometer can measure the temperature of the environment and prevent catastrophes. However, it is known that the thermometer does not prevent catastrophes, but it does provide information that can be interpreted by researchers, making it possible to predict some kind of event.

"The thermometer is very useful because we don't need to go to hospital to see if we have a fever. It can be used to see if the water isn't too hot. It can measure how hot the weather is and thus prevent some catastrophes."

In the transcript of the answer below, the students bring up an interesting argument when they express the perception that touch is not reliable for knowing if a person has a fever, since this sense depends on the temperature of the person touching and the temperature of the person being touched.

"The thermometer is important, because you can't trust your sense of touch, you can't tell if someone has a fever just by touching them; you may well have been walking in the sun, and it's hot, and they think you have a fever or not."

3.3.2.4. Fourth statement: Lavoisier and taxes

Analyzing the students' answers, we found the following ideas about the fourth statement of the second SA, which showed Lavoisier's importance as a scientist, although this didn't get him off the hook for being a tax collector. The students were asked to comment on tax collection today and to point out the positive and negative aspects of investing the money collected.

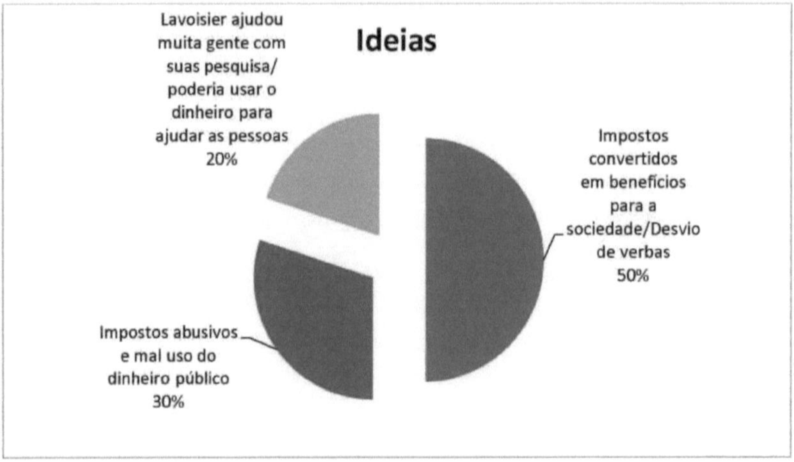

Figure 18 - Students' answers to the fourth statement of the second SA.

In this statement, the answers expressed society's feelings when it comes to tax collection, and this subject is then linked to politics and corruption. Half of the students expressed the idea that the positive point of collecting taxes is that they are converted into benefits for society, while the negative points were associated with the misappropriation of funds. 30% of the students only expressed the idea that taxes in our country are abusive and misused. 20% of the students were unable to interpret the statement and talked about Lavoisier and the taxes he charged.

In the transcript of the answers below, the students believe that the taxes charged in Brazil are high and that the money collected is not invested. However, it is known that taxes collected are invested in social benefits, not in the way the population would like, but they are invested. In the first and second transcripts below, the students place the governor and mayor as responsible for the inadequate investment of the taxes collected, as well as the detour of this money for their own benefit.

"Well, the taxes currently collected in Brazil are absurd because things are so expensive and we don't see anything happening, we expect a lot from our money but the governors deceive us."

"The positive thing is that taxes are used for sanitation, health centers and so on, and the negative thing is that the mayors are stealing all the taxes."

"Taxes are being levied on everything we buy. The money collected is not invested in anything"

In the transcript of the answers below, the students express that they know how the dynamics of tax collection work, at least in theory, and how this is converted into resources for social investment.

"Negative point: when we pay or buy something, we pay a little more tax than we should. Positive point: with the tax money we receive resources for our society."

Yes, because taxes are collected to improve our municipality and our neighborhood. After all, everything we buy or sell is taxed".

In the transcript of the answers below, the students did not interpret the statement correctly. The statement asked students to comment on tax collection today, however, students commented on Lavoisier and tax collection in Lavoisier's time and how he supposedly used the money from tax collection.

"Lavoisier was arrested and ended up literally losing his mind, and he was guillotined not for his science, but for being a private tax collector and there was nothing else tested at the time, the positive points are: that he helped many people with his research, but his negativity was simply that he was arrested at the time."

"Its positive points were the great help that Lavoisier's discovery was for people. The downside is that the money could have been used to help the people."

3.3.2.5. Fifth statement: nature of the cold

Analyzing the students' answers, the following ideas were identified with regard to the fifth statement of the second SA, in which the students had to argue whether cold is currently understood as a substance or whether it is an effect of the movement of particles.

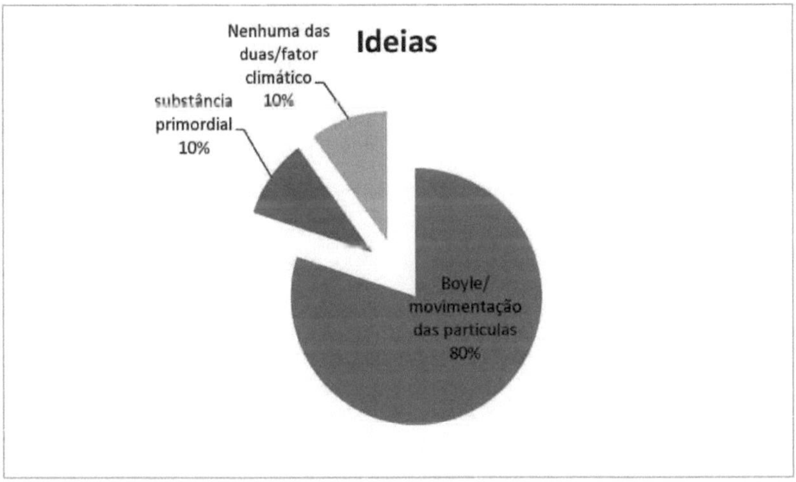

Figure 19 - Students' response to the fifth statement of the second SA

In this statement, 80% of the students say that Boyle's ideas about cold are currently accepted. However, when justifying their answers, they come up with ideas that diverge from current scientific understanding and the ideas set out in the SA text.

In the transcript below you can see that the students attribute the origin of cold to the production of foreign particles, ideas that are not attributed to Boyle in the SA text. However, it is now known that no particles are produced when cooling occurs, it is just understood that these particles move more slowly. This answer shows that the students were unable to interpret the statement and relate it to the ideas in the SA text.

"In our opinion, Boyle may be completely right, because his vision may be right, the production of strange particles that gave rise to the cold."

One group of students justified their answer by saying that Boyle's view is the correct one "*because he did experiments to prove it*". This answer highlights the view that these students have of science, giving the idea that because a scientist has done an experiment, this knowledge is true. However, we know that experiments are valid or not depending on the era in which they are carried out and according to the models and theories in which they are inserted. As in the example of Lavoisier and the caloric theory, in his time this theory was difficult to refute, but today this idea is easily refuted and even more so with the concept of energy as established by the laws of thermodynamics.

In the transcript of the answer below, the students say that neither view is correct. They reply that cold is a climatic factor due to the action and reaction of nature. This answer shows that the students were unable to interpret the text.

"In my opinion neither view is correct, the cold is a climatic factor due to the actions and reactions of nature."

In the transcript of the answer below, the students answered that the correct view is Boyle's and justified their answer by saying that cold arises when the particles of substances move, because they move and come together very slowly, causing cold. However, it is now known that this movement of the molecules generates higher temperatures and that the opposite situation, when the molecules begin to stop moving, generates lower temperatures, which is when we feel cold.

"Boyle is right, because the cold can't be seen, and we hardly understand how it is formed, nor what it is direct, the cold appears when the particles of substances move, because they come together and separate, very slowly causing the cold."

In the transcript of the answer below, the students state that the correct view is Boyle's and justify it by answering that molecules are responsible for many things, including cold and heat, showing an answer that is not in line with current scientific concepts, since we know that temperature is the measure of the kinetic energy of a body's molecules, when the temperature is higher, the molecules have more kinetic energy than when the temperature is lower, and heat is just the energy in transit that goes from a body with a higher temperature to a body with a lower temperature (PERUZZO and CANTO, 2006); (SANTOS and MOL, 2013); (FONSECA, 2013); (BONJORNO et.al, 1999); (SANT'ANNA et.al, 2010); (MORTIMER and MACHADO, 2013), so molecules are not responsible for heat.

"I believe in Boyle's thesis, because molecules are responsible for many things, including cold and heat."

In the transcript of the answer below, the students state that cold is a substance, an idea that diverges from current scientific understanding and from the ideas in the SA text.

"I think the first because everything that man says that cold was a primordial substance is also part of strange particles that gave rise to cold."

In the transcript of the answer below, the students demonstrate a certain understanding of the statement and the SA text. However, in the last two lines of the answer, the students confusingly state, in disagreement with the SA text, that Boyle thought the idea that cold gave rise to expansion was wrong. In the SA text there was no such idea that Boyle disagreed with.

"Robert Boyle's second vision was that he carried out various experiments and came to the conclusion that cold is not a substance, but something that is happening to individual particles.

which Boyle carried out several experiments, because he thought that this understanding of the nature of the cold that would give rise to expansion was wrong."

In the transcript of the answer below, the students say that when it gets cold, the particles contract. However, it is known that some substances contract when cooled and others expand, as is the case with water, so the drop in temperature is directly related to the decrease in particle movement and not to the contraction of the particles.

"The effect of the movement of the particles, because in the cold the particles contract."

3.3.2.6. Sixth statement: Lavoisier's caloric theory

Analyzing the students' answers, the following ideas were identified with regard to the sixth statement of the second SA, in which the students had to say whether the ideas of Lavoisier's caloric theory are still accepted by current science, and they had to comment on their answers.

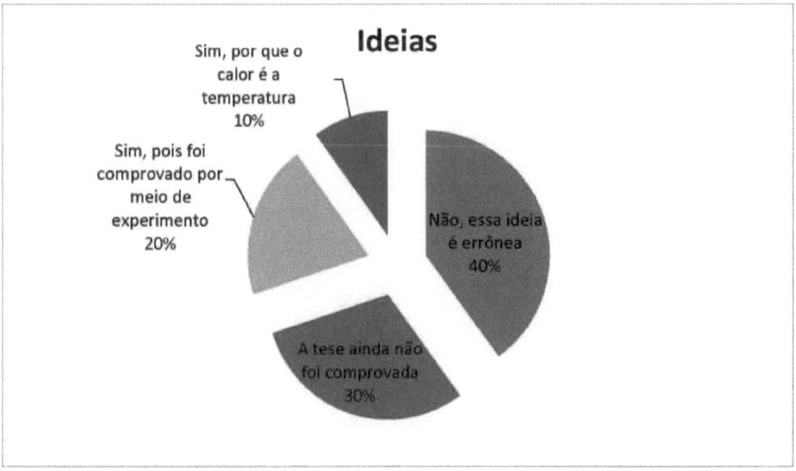

Figure 20 - Students' answers to the sixth statement of the second SA.

It was expected that most students would answer that the ideas of Lavoisier's caloric theory are not accepted by science today, since in statement 5 most students answered that the correct idea was Boyle's, which interprets cold as being the effect of the movement of the particles of a substance. However, some students contradicted themselves by stating that the ideas of Lavoisier's caloric theory were still valid for current science.

None of the students who answered no to the statement justified their answer in terms of current science; they used arguments contained in the SA text and argued about events from the time of Lavoisier and Rumford.

In the transcript of the answer below, the students clearly separate physics from science, as if physics were not part of science, nor do they argue about the ideas of caloric theory and the ideas of current science. This is evidence of an inadequate understanding of what science is.

"Yes, because today physics and science is what comments most about heat and temperature."

In the transcript of the answer below, the students present a satisfactory answer according to the information contained in the SA text, and their arguments are merely historical accounts of the SA text.

"No. Why did Lavoisier see caloric as a substance, to the point of including caloric in his list of elements? Count Rumford, for his part, was convinced that Lavoisier was wrong, and Lavoisier's story about caloric was soon undermined. But Lavoisier's theory of caloric remained dominant until the end of the 18th century".

In the transcript of the answer below, the students created information that did not appear in the SA text, stating that Rumford was more powerful and no one would have the courage to "*put another theory into practice*", however they did not relate the information in the text to the statement and to current science.

"No. Because Rumford took all the chances that his theory would be right, since Rumford was more 'powerful', no one would have the courage to put forward another empirical theory."

In the transcript of the answers below, the students believe that the ideas of Lavoisier's calories are still valid in today's scientific understanding. These ideas are at odds with the ideas contained in the SA text and with current scientific ideas, which shows that these students were unable to interpret the ideas in the SA text.

"Yes, because there are scientists who still carry out experiments that generate heat, chemical reactions in order to know how much heat there is"

In the transcriptions of the answers below, the students express ideas that do not appear in the SA text. These students were unable to interpret the ideas contained in the SA text.

"Because Lavoisier dedicated himself entirely to his work, which was the thesis of the caloric theory. Although it was contested at first, it's a good source for today's scientists."

"Why haven't experiments been carried out to prove this thesis yet, current science requires concrete proof."

"Not all because his prestige as a chemist meant that few dared to challenge his ideas."

"Because it wasn't convincing enough to make theory."

3.3.2.7. Seventh statement: technology for military purposes

Analyzing the students' answers, the following ideas were identified with regard to the seventh statement of the second SA, which contained the following information: "*Count Rumford was responsible for artillery and from his experiments with gun drilling he had enough arguments to challenge the caloric theory. Throughout the history of science, many scientific and technological advances have been achieved through the military needs of guerrilla warfare, the best known of which is the atomic bomb*". Students should argue their views on the use of technology for military purposes.

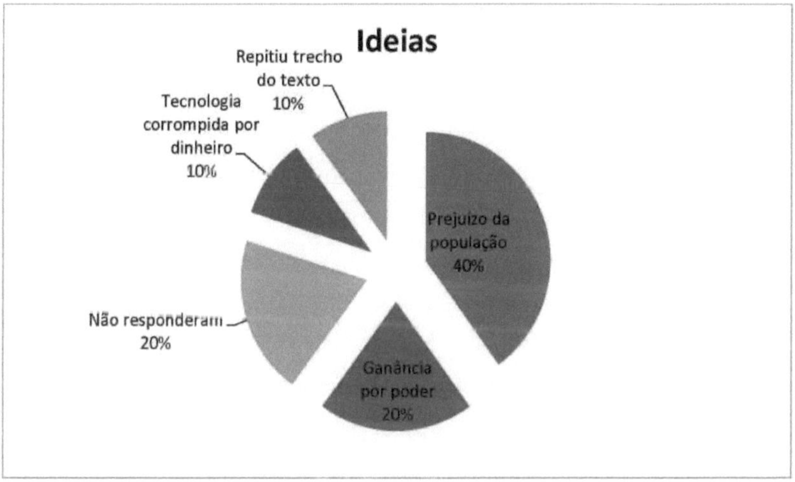

Figure 21 - Students' response to the seventh statement of the second SA

Not everyone argued about technology for military purposes, but an almost unanimous 70% of students agreed that the use and advancement of technology for war purposes would be a bad thing.

In the transcripts below, the students don't seem to understand what the statement was asking for, which was to comment on the use of science and technology for military purposes and were given the example of the atomic bombs that were dropped on Hiroshima and Nagasaki. However, these students understood that it was to argue in relation to the atomic bomb, demonstrating that they were unable to interpret the statement.

"It caused a lot of damage to the population because of the radiation and it damaged the population."

"This technology, the atomic bomb, is a threat to the world, because if there is a third world war and these bombs are dropped, the whole world will come to an end because today's atomic bombs are more technological and more threatening than those that were dropped at the end of the second world war in Yroshima and Nagazaqui. I believe that today these bombs are a threat to humanity".

In the transcripts of the answers below, the students argue that technology is corrupted by money and used for war, believing that the more technology there is, the more war there will be.

"Technology has been corrupted by money and thus used for what we believe is the most futile thing in the world - war."

"The advance of technology is very interesting, but for wars or even for the military, too much technology is no use, because if I have a lot of technology and advanced weaponry, they wouldn't want to give up the war, and with that there would never be peace."

In the transcripts of the answers below, the students report some of the information that appears in the SA text, even attributing the development of the atomic bomb to Count Rumford, showing that the students were unable to interpret the statement.

"Rumford had noticed that the friction in drilling a barrel generated a great deal of heat. So he adapted the drilling machine to produce even more heat by installing a blunt bit that had one end submerged in a water jacket."

"Count Rumford was responsible for the artillery from his experience and from the course of the history of science and many technological advances Then came the atomic bomb etc.

In the transcript of the answer below, the students argue that instead of investing the money in making bombs, it would be better to invest it in fighting child hunger in Africa, demonstrating that they are in tune with some of the everyday facts that appear in the media.

"In my opinion I don't accept it because if there was a war in our country thousands and thousands of innocent people could be killed while they invest money in bombs for war, children die of hunger in Africa."

3.3.3 Discussion of the students' ideas about the SA statements

When analyzing the contribution of the AS developed for the execution of the pedagogical moment of *"initial problematization"* for the teaching and learning process, one notes the importance of this moment, which is believed to be a very important diagnostic moment, which does not always have its place in the teaching and learning processes that take place daily in schools and universities.

Through the initial problematization, conceptual cognitive barriers can be perceived, which would not be noticed if this very important stage were neglected. In carrying out this pedagogical moment, one notices what Freire (2005) calls naive consciousness, which most of the time originates in popular language, with only ideas and concepts that are passed down from generation to generation.

With this evidence described in the previous paragraph, we believe that the *pedagogical moments* and the *conceptual profiles* can complement each other and contribute even more to the teaching and learning processes, where we believe that the student is the active subject and participant in their own learning.

With the application of SA, 25 conceptual cognitive barriers exposed by the students were identified, which will be presented below:

1. Cold is a substance;
2. Cold is a reaction of the body;
3. Cold is only related to humid climates;
4. Cold is a drop in temperature;
5. Cold is the lack of heat in a body;
6. Cold is an air mass;
7. Cold is the change in climate to wetter or cloudier weather;
8. Cold weather causes illnesses such as flu, leptospirosis and others;
9. Heat is a substance;
10. Temperature is the measure of a body's heat;
11. Saving money is just cutting costs;
12. Volcanoes evaporate in cold seasons,
13. Volcanic lava is temperature;
14. Summer only exists if it's hot;
15. Rainy seasons are bad for crops;
16. Weather is synonymous with climate;
17. Thermometer controls the temperature;
18. Thermometer evaluates health status;
19. Mercury is several types of substance;
20. Mercury is red;
21. Knowing the temperature of the environment can prevent disasters;
22. If a scientist conducts experiments, his theory is true;
23. Scientists today carry out experiments to measure how many calories are in a body;
24. Physics is not part of science;

25. Count Rumford began the first studies of the atomic bomb.

These conceptual cognitive barriers show the distance between common sense concepts and scientific concepts, which should be taken into account by teachers when designing their lessons.

After applying the *initial problematization*, the teacher is able to attack (in a good way) the students' previous conceptions (common sense) that are diverging from the scientific conceptions. However, we want to emphasize that the intention is not to destroy the students' previous conceptions, but to introduce them to other (scientific) ways of thinking and conceiving the world we live in.

The 25 conceptual cognitive barriers identified can be classified into 4 categories:

- In the first category are the conceptual cognitive barriers from 1 to 10, which are associated with the ideas of cold, heat and temperature.
- In the second category are conceptual cognitive barriers 12, 13, 14, 15, 16 and 21, which are associated with ideas of natural phenomena.
- In the third category are conceptual cognitive barriers 11, 17, 18, 19 and 20, which are associated with everyday ideas.
- In the fourth category are conceptual cognitive barriers 22 to 25, which are associated with ideas about how science works.

As our main interest is the ideas related to the concepts of heat and temperature, we will analyze the first category, which concerns the ideas of cold, heat and temperature.

In the article by Amaral and Mortimer (2001), the authors establish the conceptual profile of heat and divide it into 5 zones: Animist, Substantialist, Realist, Empiricist and Rationalist. The conceptual cognitive barriers identified in this work on the concepts of cold, heat and temperature can be classified as belonging to the Animist, Substantialist and Realist zones of Amaral and Mortimer (2001), as shown in Table 1.

Table 1 - Conceptual cognitive barriers belonging to the conceptual profile zones

Conceptual Profile Zones	Conceptual Cognitive Barriers
Animist	8
Substantialist	1, 5, 6, 9 e 10
Realistic	2, 3, 4 e 7

Thus, we believe that after this first moment *(initial problematization)* some ideas from the *conceptual profiles* can be used to make students aware of the need to acquire new knowledge, but that they don't need to abandon the knowledge they already have, and that they can live with both types of knowledge and use one or the other depending on the need.

3.4 Actions to implement the SA

3.4.1. Cell phone use in the classroom

The abusive and inappropriate use of cell phones in the classroom was a very uncomfortable event that caught our attention. It is believed that the cell phone or any other type of ICT when used for research purposes or activities related to the subject of the lesson and with well-planned practices by the teacher is a good pedagogical tool, but most of the time this technological tool is only used to access social networks, take photos, send messages and photos, paste in assessments.

So how are students supposed to participate as active subjects in their own learning if most of the time they are disconnected from the class in a virtual world?

When researching the use of cell phones in the classroom, we found Law 8.949 of November 2009 (available in Annex 1), which establishes a ban on the use of cell phones in classrooms in public and private schools in the state of Paraiba.

Law 8.949 is not so new, it has been in force for more or less 5 years and what we see in classrooms is its increasingly frequent use for non-didactic purposes, a fact that hinders teaching practices and is a source of discouragement for elementary school teachers in the final grades and especially secondary school teachers.

3.4.2. Students' textual production

In this section we will discuss the students' textual production, which was another point that attracted a lot of attention after the SA was applied. In section 3.3, we presented verbatim transcriptions of the students' answers, with the aim of disturbing the reader with some very crude sentences that show a high degree of disagreement with current spelling norms. All the students' answers are included in Appendix 5 for the reader's appreciation.

After analyzing the students' responses to the SA, a report containing the students' responses to the SA was sent to the Portuguese teachers and the school council for their appreciation and discussion of possible solutions.

In the school planning for the month of May, the subject of textual production was put on the agenda and discussed. All the teachers were unanimous in recognizing the deficiencies of secondary school students. The teachers reported that in secondary school, which is the last stage of basic education, there is not much to do, as this would be a problem that should be worked on more rigorously in elementary school.

This preoccupation with students' textual production seems not to be limited to the high school students at this school. According to the education website G1 of the Rede Globo broadcaster, 52,974 students taking part in Enem 2014 scored zero in the essay test, which represents around 8.5 of the candidates taking part in Enem 2014.

This fact, which occurred in the 2014 Enem, shows that the deficiency in textual production is not just among the students at the Immaculada Conceiçao School in Cabedelo, but that this level of deficiency is alarming at a national level.

3.5 Students' answers to the pre-test

This pre-test was administered the next day after the SA and is available in item 4 of the Teaching Situation (SE). It is an assessment of the students' preconceptions about the topic of heat and temperature. Some of the questions in this pre-test put students in situations that they can experience in their daily lives.

The SE is designed to help the teacher, guiding their work and their attitude in the classroom. It includes the activities that will be developed in the classroom, as well as complementary concepts of the subject, possible answers to the activities in the classroom, assessments and documents that are based on the teaching of the subject in question.

In SE the questions do not appear in the order that will be shown in sessions 3.5.1 and 3.5.2. The questions are divided into two groups of concepts. The first group that will be shown is related to the concept of heat, while the second group presented is related to the concept of temperature and thermal equilibrium.

The questions will be shown with the multiple choice alternatives that the student could choose from. The students' answers are shown in a table containing the alternative, the number of students who answered this alternative and the percentage of students who answered the alternative. For the reader's better appreciation, the correct alternative for each question is highlighted in yellow in each answer box.

3.5.1. Students' answers to the pre-test on the concept of heat

01-A person says that their blanket is good "because it stops the cold coming through". This statement is true:

a) true

b) fake

c) depends on the material of the blanket

d) depends on which country you're in

e) it's true if the air isn't warmer than the blanket

Alternatives	No. of students	%
a)	20	50
b)	2	5
c)	9	22,5
d)	5	12,5
e)	4	10

02-(CEFET-SP) Heat is:

a) energy in transit from one body to another when *there is a* difference in temperature between them.

b) measured in degrees Celsius.

c) a form of energy that does not exist in cold bodies.

d) a form of energy that is attributed to hot bodies.

e) the same as temperature.

Student response

Alternatives	No. of students	%
a)	4	10
b)	9	22,5
c)	12	30
d)	13	32,5
e)	2	5

03-(PUCCAMP-SP) Regarding the concept of heat, it can be said that it is a:

a) measuring the temperature of the system.

b) form of energy in transit.

c) fluid substance.

d) quantity related to friction.

e) energy that bodies possess.

Student response

Alternatives	No. of students	%
a)	9	22,5
b)	4	10
c)	3	7,5
d)	8	20
e)	16	40

04-(UFP-RS) Consider the following statements:

I. When two bodies are in thermal equilibrium, they both have the same amount of heat.

II. When two bodies are in thermal equilibrium, they both have the same temperature.

III. Heat is the transfer of temperature from one body to another.

IV. Heat is a form of energy in transit.

From the above statements, it can be said that:

a) I, II, III and IV are correct.

b) I, II, III are correct.

c) I, II, IV are correct.

d) II and IV are correct.

e) II and III are correct.

Student response

Alternatives	No. of students	%
a)	16	40
b)	14	35
c)	2	5
d)	3	7,5
e)	5	12,5

05-(OSEC-SP) The fact that heat passes from one body to another is due to:

a) amount of heat in each one.

b) temperature difference between them.

c) total kinetic energy of its molecules.

d) the number of calories in each one.

e) none of the above is true.

Student response

Alternatives	No. of students	%
a)	13	32,5
b)	3	7,5
c)	6	15
d)	7	17,5
e)	11	27,5

06-(PUC-MG) If heat is exchanged between two bodies, it is correct to say that, at the beginning of this process, they are different:

a) Your pasta.

b) Their thermal capacities.

c) Their specific heats.

d) Their temperatures.

e) Their energy.

Student response

Alternatives	No. of students	%
a)	8	20
b)	8	20
c)	7	17,5
d)	6	15
e)	11	27,5

07-(UNIFESP-SP) The SI (International System of Units) uses the joule as the unit of heat, because heat is energy. However, it only makes sense to talk about heat as energy in transit, i.e. energy that is transferred from one body to another as a result of the difference in temperature between them.

Choose the statement in which the concept of heat is used correctly.

a) The temperature of a body decreases when it loses some of the heat that was stored in it.

b) A body's temperature rises when it accumulates heat.

c) The temperature of a body decreases when it gives off heat to the environment.

d) An increase in the temperature of a body is an indicator that it has stored heat.

e) A body can only reach absolute zero if it is emptied of all the heat it contains.

Student response

Alternatives	No. of students	%
a)	9	22,5
b)	6	15
c)	12	30
d)	7	17,5
e)	6	15

08-(UEPB-PB) In a Physics lesson, a student is asked to explain physically what happens when a piece of hot iron is placed in a container of cold water. He says: "The iron is hot because it contains a lot of heat. The water is colder than the iron because it contains less heat than the iron. When the two are placed together, part of the heat contained in the iron passes into the water, until they have the same level of heat... and then they are in equilibrium." Taking the student's statements as a reference and considering the scientifically correct concepts, analyze the following propositions:

I. According to the current concept of heat, the expression: "The iron is hot because it contains a lot of heat" is wrong.

II. Instead of stating: "... part of the heat contained in the iron passes into the water", the student should say that "there is a transfer of temperature between them".

III. "... until they have the same level of heat... and then they reach equilibrium", correct, because when two bodies reach thermal equilibrium their specific heats equalize.

Choose the correct alternative:

a) All propositions are true.

b) Only proposition I is true.

c) Only proposition II is true.

d) Only proposition III is true.

e) Only propositions I and III are true.

Student response

Alternatives	No. of students	%
a)	11	27,5
b)	6	15
c)	4	10
d)	10	25
e)	9	22,5

09-(UFRGS) For two bodies to be able to exchange heat, it is necessary that:

I - are at different temperatures.

II - have different masses.

III - there is a heat-conducting medium between them.

Which of these statements is (are) correct?

a) only I.

b) only II.

c) only I and II.

d) only I and III.

e) all of them.

Student response

Alternatives	No. of students	%
a)	9	22,5
b)	7	17,5
c)	9	22,5
d)	13	32,5
e)	2	5

Analyzing the students' answers, the following conceptual cognitive barriers were identified:

1. Cold as a substance;

2. Heat as some kind of energy (or something) that exists in hot bodies and does not exist or exists little in cold bodies;

3. Unfamiliarity with the concepts of heat and temperature;

4. Heat as a substance;

5. Ignorance of the conditions necessary for heat transfer between bodies;

6. Believe that there must be a heat-conducting medium between bodies so that one body can transfer heat to another.

3.5.2. Students' answers to the pre-test on the concept of temperature and thermal equilibrium

Ol-(UFV-MG) When two bodies made of different materials are in thermal equilibrium, isolated from the environment, it can be said that:

a) the hottest is the one with the least mass.

b) despite the contact, their temperatures do not vary.

c) the warmer one provides heat to the colder one.

d) the colder one provides heat to the warmer one.

e) their temperatures depend on their densities.

Alternatives	No. of students	%
a)	11	27,5
b)	7	17,5
c)	14	35
d)	3	7,5
e)	5	12,5

02-(MACKENZIE-SP) The famous Irish physicist William Thomson, who became known worldwide as Lord Kelvin. Among his many works, he "created" the absolute thermometric scale. This scale, known as the Kelvin scale, consequently does not allow negative values and, to this end, established the state of molecular rest as zero. Conceptually, his position is consistent, since the temperature of a body refers to the measurement:

a) of the amount of movement of the body's molecules.

b) the amount of heat in the body.

c) of the thermal energy associated with the body.

d) the thermal energy of the body's molecules.

e) the degree of agitation of the body's molecules.

Student response

Alternatives	No. of students	%
a)	8	20
b)	14	35
c)	4	10
d)	9	22,5
e)	5	12,5

03-(FEI-SP) A system thermally isolated from the environment has three bodies, one made of iron, one of aluminum and one of copper. After a certain period of time, the temperatures of the iron and aluminium have increased. We can conclude that:

a) the copper body has also increased its temperature.

b) the copper body gained heat from the aluminum body and gave heat to the iron body. c) the copper body gave heat to the aluminum body and received heat from the iron body. d) the copper body remains at the same temperature.

e) the copper body has lowered its temperature.

Student response

Alternatives	No. of students	%
a)	7	17,5
b)	9	22,5
c)	8	20
d)	9	22,5
e)	7	17,5

04- (FATEC-SP) A system A is in thermal equilibrium with another system B and the latter is not in thermal equilibrium with another system C. So, we can say that:

a) systems A and C have the same amount of heat.

b) the temperature of A is different from that of B.

c) systems A and B have the same temperature.

d) the temperature of B is different from that of C, but C can have the same temperature as system A.

e) none of the above.

Student response

Alternatives	No. of students	%
a)	6	15
b)	9	22,5
c)	6	15
d)	12	30
e)	7	17,5

05-(UFSCAR-SP) Two bodies A and B, of masses mA and mB, are initially at temperatures tA and tB, respectively, with tA≠ tB. At a given moment, they are brought into thermal contact. After reaching thermal equilibrium, we will have:

a) $t'_A > t'_B$

b) $t'_A < t'_B$ c) $t'_A = t'_B$ d) mA=mB e) n.a.d.

Student response

Alternatives	No. of students	%
a)	12	30
b)	10	25
c)	10	25
d)	3	7,5
e)	5	12.5

06- A thermometer is enclosed in a glass bulb where a vacuum is created. Suppose that the vacuum is perfect and that the thermometer is measuring the room temperature, 25°C. After a while, the room temperature rises to 30°C. It is then observed that the thermometer is marked.

a) also rises, and tends to reach thermal equilibrium with the environment.

b) is kept at 25°C, whatever the ambient temperature.

c) tends to decrease continuously, regardless of the ambient temperature.

d) will rise, but never reach thermal equilibrium with the environment.

e) tends to reach the minimum value on the thermometer scale.

Student response

Alternatives	No. of students	%
a)	7	17,5
b)	14	35
c)	1	2,5
d)	10	25
e)	8	20

07-(UNESP-SP) When a nurse places a mercury thermometer under a patient's tongue, for example, she always waits a while before taking a reading. This time interval is necessary.

a) so that the thermometer comes into thermal equilibrium with the patient's body.

b) so that the mercury, which is very heavy, can go up the capillary tube.

c) so that the mercury passes through the bottleneck in the capillary tube.

d) due to the difference between the specific heat values of mercury and the human body.

e) because the coefficient of expansion of glass is different from the coefficient of expansion of mercury.

Student response

Alternatives	No. of students	%
a)	8	20
b)	7	17,5
c)	6	15
d)	10	25
e)	9	22,5

08- (CFT-SC-010) In our homes, we usually use wooden or rubber flooring in the bedrooms and ceramic flooring in the kitchen. Why do ceramic floors feel colder?

a) Because ceramic floors are warmer than wooden floors, that's why ceramic floors feel colder.

b) Because ceramic floors are colder than wooden floors, which is why ceramic floors feel colder.

c) Because the ceramic floor in the bedroom gives it a less elegant tone.

d) Because wooden floors exchange less heat with our feet, causing us to feel less cold.

e) Because ceramic flooring has more contact area with the foot, it exchanges more heat, causing us to feel cold.

Student response

Alternatives	No. of students	%
a)	3	7,5
b)	19	47,5
c)	1	2,5
d)	7	17,5
e)	10	25

Analyzing the students' answers, the following conceptual cognitive barriers were identified:

7. You don't know the concept of thermal equilibrium;
8. Temperature is the measure of the amount of heat in a body;
9. The sensations of hot and cold are equivalent to the concepts of heat and temperature.

3.5.3 Analysis of pre-test answers

After analyzing the data obtained in sections 3.5.1 and 3.5.2, it was found that most students have the idea that heat is a substance, like something inside bodies, especially those that are hot.

Another idea very present among the students is that temperature is the measure of heat that exists in bodies and bodies that are below zero degrees have no heat. It is also clear that the students are not aware of the concept of thermal equilibrium.

These ideas expressed by the students are in line with the literature (ZYLBERSZTAJN, 1983; SILVA, 1995; MORTIMER and AMARAL, 1998; AMARAL and MORTIMER 2001; GRINGS, CABALLERO and MOREIRA, 2006, 2007 and 2008; GONÇALVES, 2005; RAFAEL, 2007; SILVA Jr, 2007).

With the application of the pre-test, 9 conceptual cognitive barriers were identified, of which barrier 1 and 4 were also identified in the students' responses regarding SA, only conceptual cognitive barrier

1 does not appear in the literature cited in this paper. The ideas of heat and temperature are widely discussed in the literature. The ideas of cold are little discussed, so the ideas of cold as a substance do not appear in the literature (or appear little). For this reason, we risk placing the conceptual cognitive barrier of cold as a substance as being poorly diagnosed (or undiagnosed) in the literature.

Associating the conceptual cognitive barriers identified in the students' answers to the pre-test with the zones of Amaral and Mortimer's (2001) conceptual profile for heat, it was shown that the conceptual cognitive barriers belong to the substantialist and realistic zones of the conceptual profile of heat, according to Table 2.

Table 2 - Conceptual cognitive barriers belonging to the conceptual profile zones

Conceptual Profile Zones	Conceptual Cognitive Barriers
Substantialist	1, 4 e 8
Realistic	2, 3, 5, 6, 7 e 9

These analyses can be used as a means of making students aware of the need to acquire new knowledge, which better adapts to and explains the phenomena that happen around us on a daily basis, so that they can consciously use this new knowledge for the benefit of society.

CHAPTER IV: FINAL CONSIDERATIONS

This research revealed the importance of *initial problematization.* Among the many contributions of this *pedagogical moment*, the most characteristic is that of diagnosing students' previous conceptions (world knowledge) about a given topic. We are also able to identify barriers at a conceptual level, and plan practices with the aim of overcoming these barriers (without destroying the world knowledge brought by the students). From this diagnosis, the teacher is able to carry out a problematizing practice, placing the students as the subjects of their own learning.

With the application of the *initial problematization*, it becomes clear that care must be taken in the social, cultural, historical, everyday and geographical approaches that will be used to address the topic of interest, because the common sense knowledge rooted in most people is very strong and cannot be confronted directly.

The aim is to make students aware of the need for new, more comprehensive (scientific) knowledge that can better describe the phenomena that occur in the world in which they live. It is at this point that we believe that the ideas of *conceptual profiles* can complement *teaching moments.*

It is believed that new (scientific) knowledge should be used to explain situations that world knowledge (common sense) cannot explain, trying to enable students themselves to realize their conceptual cognitive barriers and feel the need to overcome them, as well as to use more plausible and complete knowledge.

With these practices, the student must acquire the information and synthesize it into knowledge, aware that common sense language is necessary in everyday communication, but that in a formal setting such as a classroom or scientific event, it is advisable to use scientific language.

The teacher must therefore develop practices that offer the conditions for students to become aware of their existence and, as such, make conscious decisions to bring about a positive change in their *status quo*. In this sense, there is a need to propose teaching that broadens the areas of the student's *conceptual profiles*, who will have to acquire new (scientific) understandings, which will begin to coexist with the old (everyday) understandings.

Considering that we live in a highly technological world in which information spreads across the planet in a matter of seconds, one of the aims of teaching chemistry, especially at secondary school level, is to enable students to (de)code information and act in a conscious way as a thinking, critical and productive citizen in the world in which they live.

This requires a change in the teacher's attitude towards the methodology to be used in the classroom, which includes, among other factors, the production or adaptation of the teaching material to be used.

This change must be flexible and accompany social transformations, understanding the individual in their full formation. Teachers must develop the skills in their students that will enable them to become agents of social transformation.

We believe in the emergence of reflective teaching practices that are less rigid in their defense of absolute truths and immutable certainties, which appear with epistemological and ontological depth in the words of Japiassu (1999, p.328) who says: "*A frozen truth becomes an intellectual anesthesia. Its paralyzing effect generates countless diseases of the spirit, including adult paralysis of intelligence*".

As we don't like conclusions, because they are finished ideas, we don't believe that this work is the end of a thought, but a guide to new directions that will help us learn, understanding ourselves beyond what our eyes can see, without being afraid of uncertainties, because they represent the obstacles that motivate us to seek intellectual development and critical thinking.

REFERENCES

- ALBERT, E. Development of Concept of Heat in Children. *Science Education.* p. 389-399, 1978.

- ALVES-MAZZOTTI, A. J.; GEWANDSZNAJDER, Fernando. The method in the natural and social sciences: quantitative and qualitative research. 2. ed. Sao Paulo: Pioneira, 1999.

- ALVETTI, M. A. S.; DELIZOICOV, D. Teaching modern and contemporary physics and the magazine Ciência Hoje. In: Encontro de Pesquisa em Ensino de Fisica, 1998, Florianópolis. Abstracts. Florianópolis: Brazilian Physics Society, 1998.

- AMARAL, E. M. R.; MORTIMER, E. F. A proposal for a conceptual profile for the concept of heat. *Rev. brasileira de pesquisa em educaçdo em ciências*. n. 3. P. 5-18. 2001, v.1.

- ANGOTTI, J. A. P. Alternative solution for the training of science teachers: an educational project developed in Guinea Bissau. Dissertation (Masters in Education) - Faculty of Education, University of Sao Paulo, Sao Paulo, 1982.

- ATKINS, P; JONES, L. Principles of chemistry: questioning life and the environment (transl. Ricardo Bicca de Alencastro). 3. ed. Porto Alegre: Bookman, 2006.

- AUTH, M. A. et al. Situação de Estudo na área de ciências do ensino mèdio: rompendo fronteiras disciplinar. In: MORAES, R.; MANCUSO, R. (Org.). Science education: curriculum production and teacher training. Ijui: Editora Unijui, p. 253-276, 2004.

- BACHELARD, G. *The Philosophy of Nothing*. The Thinkers Collection. Sao Paulo: Abril Cultural. 1978.

- BACHELARD, G. *The Formation of the Scientific Spirit* (transl. Estela dos Santos Abreu). Rio de Janeiro: Contraponto. 1996.

- BASSALO, J. M. F. The History of Heat: Thermometry. *Rev. de Ensino de Fisica.* n.1. p. 135-161. 1991, v.13.

- BASSALO, J. M. F. Crònicas da Fisica. Belém: UFPA. 1992.

- BARBOSA LIMA, M. C. e LINS DE BARROS, H. Uma proposta de ensino de calor e temperatura à luz de Bachelard. *Proceedings of the 1st National Meeting of Research in Science Teaching*, p. 315-321. 1997.

- BARRETO, R. G. Discursos, Tecnologias, Educaçao. Rio de Janeiro: EdUERJ (Sèrie Pesquisa em Educaçao, Pràticas de Linguagem), 2009

• BAUER, M. W.; GASKELL, G. (Orgs.). Qualitative research with text, images and sound. Manual pràtico. Petrópolis: Vozes, 2002.

• BONJORNO, R. G.; CLINTON, M. Fisica Fundamental. Sao Paulo: FTD, 1999.

• BOSS, S. L. B. et al. Teaching by investigation: report of a pedagogical experience in thermodynamics. In: XVIII National Symposium on Physics Teaching, 2009. Vitória. Proceedings of the XVIII National Symposium on Physics Teaching. Vitória: 2009.

• BRASIL. National Curriculum Parameters for Secondary Education. Brasilia: MEC/SEMTEC, 1999.

• BRAZIL SCHOOL. Influenza. Available at <http://www.brasilescola.com/doencas/gripe.htm>. Accessed January 2, 2015.

• BROOK, A.; BRIGGS, H.; BELL, B. E; DRIVER, R. Aspects of Secondary Students' Understanding of Heat: Summary Report. (CLIS Project - Children's Learning in Science Project) Published by the University of Leeds. 1984.

• CAREGNATO, R. C. A.; MUTTI, R. Qualitative research: discourse analysis versus content analysis. Florianópolis: Texto Contexto Enfermagem, 2006.

• CARVALHO, A. M. P. et al. Termodinàmica: Um ensino por investigação. 1. ed. Sao Paulo: University of Sao Paulo - Faculty of Education, 1999, v.1.

• CERVANTES, A. Los Conceptos de Calor y Temperatura: Una revision Bibliografica. *Teaching Science*, p. 66-70. 1987.

• CINDRA, J. L. TEIXEIRA, O. P. B. Heat and temperature and their implications through a historical approach. In: MARTINS, R. A. MARTINS, L. A. C. P. SILVA. C. C. Ferreira, J. M. H. (eds.). Philosophy and history of science in the Southern Cone: 3rd meeting. Campinas: AFHIC, p. 240-248, 2004.

• COBERN, W. W. Worldview theory and conceptual change in science education. Science Education, p. 579-610, 1996.

• COMBIE, A. C. Historia de la ciencia: de San Agustin a Galileo (translated by J. Bernia). 4. ed. Madrid: Alianza Editorial, 1985, v.2.

• DELIZOICOV, D. Concepçao problematizadora para o ensino de ciências na educação formal: relato e análise de uma pràtica educacional na Guiné-Bissau. Dissertation (Master's Degree in Science Teaching) - Institute of Physics, University of Sao Paulo, Sao Paulo, 1982.

• DELIZOICOV, D. Knowledge, tensions and transitions. 1991. Thesis (Doctorate in Education) -

Faculty of Education, University of Sao Paulo, Sao Paulo, 1991.

- DELIZOICOV, D. Problems and problematizations. In: PIETROCOLA, M. (Org.). Teaching physics: content, methodology and epistemology in an integrative conception. Florianópolis: Ed. da UFSC. p. 125-150, 2001

- DELIZOICOV, D. La educación en ciencias y la perspectiva de Paulo Freire. Alexandria: revista de educaçao em ciência e tecnologia, Florianópolis, v. 1, n. 2, p. 37-62, 2008. Available at: <http://www.ppgect.ufsc.br/alexandriarevista/numero_2/artigos/demetrio.pdf>. Accessed on: 10/12/2013.

- DELIZOICOV, D.; ANGOTTI, J. A.; PERNAMBUCO, M. M. Ensino de ciências: fundamentos e métodos. Sao Paulo: Cortez, 2002.

- DEMO, P. Avaliaçao qualitativa. Sao Paulo: Cortez, 1991.

- DEMO, P. Introduçao à metodologia da ciência. 2. ed. Sao Paulo: Atlas, 1995.

- ERICKSON, G. Heat and Temperature - part A: An overview of pupils' ideas. In Drive, R.; Guesne, E. and Tiberghien, A. (eds.) *Children's Ideas in Science*. Open University Press: Milton Keynes - Philadelphia. 1985.

- EXAME, 6 reasons why 50°C heat is hellish. Available at:< http://exame.abril.com.br/mundo/noticias/5-razoes-por-que-um-calor-de-50c-pode-ser-infernal?page=1> Accessed on February 26, 2014.

- FEAUSP, What is economics. Available at:< http://www.fea.usp.br/conteudo.php?i=202> Accessed on August 20, 2014.

- FERREIRA, V. F. Interactive technologies in teaching. Quim. Nova, 1998, v.21.

- FIDEL, R. The case study method: a case study. In: GLAZIER, Jack D. & POWELL, Ronald R. Qualitative research in information management. Englewood, CO: Libraries Unlimited, p.37-50, 1992.

- FLICK, U. Uma introduçao à pesquisa qualitativa. 2. ed. Porto Alegre: Bookman, 2004.

- FONSECA, M. R. M. Quimica. 1. ed. Sao Paulo: Atica, 2013, v.2.

- FREIRE, P. Pedagogy of the oppressed. 43ª ed. Rio de Janeiro: Paz e Terra, 2005.

- FREIRE, P. Pedagogia da autonomia: saberes necessàrios à pràtica educativa. 33rd ed. Rio de Janeiro: Paz e Terra, 2006.

- G1/Educação, 529 thousand students scored zero on the Enem 2014 essay, says MEC. Available at< http://g1.globo.com/educacao/noticia/2015/01/529-mil-alunos-obtiveram- nota-zero-na-redacao-do-enem-2014-diz-mec.html> Accessed on January 15, 2015.

- G1/Fantàstico, Experts explain this summer's extreme heat in Brazil. Available at:< http://g1.globo.com/fantastico/noticia/2014/02/fantastico-explica-calor-extremo-deste- verao-no-brasil.html> Accessed February 26, 2014.
- GEHLEN, S. T.; MALDANER, O. A.; DELIZOICOV, D. Pedagogical moments and the stages of the study situation: Complementarities and contributions to Education in
- Sciences. Ciênc. educ. (Bauru) [online]. n.1, p. 1-22. 2012, v.18.
- GIL, A. C. Como elaborar projetos de pesquisa. Sao Paulo: Atlas, 1991.
- GIORDAN, M. The computer in science education: a brief critical review of some forms of use. Ciência & Educaçao, v. 11, n. 2, p. 279-304, 2005
- GIORDAN, M. Some technical and methodological questions about recording classroom action: digital capture and storage. In: SANTOS, F. M. T.; GRECA, I. M. (Org.). A pesquisa em ensino de ciências no Brasil e suas metodologias. Ijui: Editora Unijui, p. 213-238, 2006.
- GOLDENBERG, M. A arte de pesquisa. Rio de Janeiro: Record, 1997.
- GONÇALVES, L. J. Use of animations for meaningful learning of thermal physics in secondary education. Dissertation (Professional Master's Degree in Physics Teaching) - Federal University of Rio Grande do Sul, Porto Alegre, 2005.
- GRINGS, E. T. O. CABALLERO, C. MOREIRA, M. A. Possible indicators of operative invariants presented by students in thermodynamics concepts. Revista Brasileira de Ensino de Fisica, p. 1-9, 2006, v. 24
- GRINGS, E. T. O. CABALLERO, C. MOREIRA, M. A. Meanings of thermodynamic concepts and possible indicators of operational invariants presented by high school students. Revista Liberato (Novo Hamburgo), p. 7-12, 2007, v.8.
- GRINGS, E. T. O. CABALLERO, C. MOREIRA, M. A. Students' progress and setbacks in the conceptual field of thermodynamics. REEC. Revista Electrónica de Ensenanza de las ciências, p. 23-46, 2008, v.7.
- HELLER, A. O cotidiano e a história. Rio de Janeiro: Paz e Terra. 1989
- HELLER, A. Sociologia de la vida cotidiana. Barcelona: Peninsula. 1991
- HOPPE, E. *Histoire de la Pysique* (translated by H. Besson). Paris, Payot, 1928.
- JAPIASSU, H. As Paixes da Ciência: estudos de História das Ciências. 2. ed. Sao Paulo: Letras &

Letras, 1999.

- LABURU, C. E. Uma descriçao da forma do pensamento dos alunos em sala de aula. Revista Brasileira de Ensino de Fisica, n° 3, p. 243-253, 1995, v.17

- LANKSHEAR, C.; KNOBEL, M. Pedagogical research: from design to implementation. Translated by Magda França Lopes. Porto Alegre: Artmed, 2008.

- LÜDKE, Menga e ANDRÉ, Marli E. D. A. Pesquisa em educaçao: abordagens qualitativas. Sao Paulo: EPU, 1986.

- MALDANER, O. A. Situaçoes de estudo no ensino mèdio: nova compreensão de educaçao bàsica. In: NARDI, R. (Org.). Pesquisa em ensino de ciências no Brasil: alguns recortes. Sao Paulo: Escrituras, p. 237-253, 2007.

- MALDANER, O. A.; ZANON, L. B. Situaçao de Estudo: uma organização de ensino que extraprapola a formação disciplinar em ciências. In: MORAES, R.; MANCUSO, R. (Org.). Science education: curriculum production and teacher training. Ijui: Editora Unijui, p. 43-64, 2004

- MARQUES, L. R. M; ARAÙJO I. S. Fisica tèrmica. Porto Alegre: UFRGS, n.5, 73p, 2009. v.20

- MINAYO, M. C. S.; MINAYO-GOMÉZ, C. Difficult and possible relationships between quantitative and qualitative methods in the study of health problems. In: GOLDENBERG, P.; MARSIGLIA, R. M. G.; GOMES, M. H. A. (Orgs.). The classic and the new: trends, objects and approaches in social sciences and health. Rio de Janeiro: Fiocruz, p.117-42, 2003.

- MINAYO, M. C. S. O desafio do conhecimento. Qualitative research in health. Sao Paulo: HUCITEC, 2007.

- MOREIRA, D. A. The phenomenological method in research. Sao Paulo: Pioneira Thomson, 2002.

- MOREIRA, A.F.B. and KRAMER, S. Contemporaneity, education and technology. Educaçao e Sociedade, n.100 - Especial, 2007, p. 1037-1057, v. 28.

- MORTIMER, E. F. Evolution of atomism in the classroom: Changing conceptual profiles. Thesis (Doctorate in Education) - Faculty of Education, University of Sao Paulo, Sao Paulo, 1994.

- MORTIMER, E. F. Conceptual change or conceptual profile change? Science & Education, 4, 265-287, 1995.

- MORTIMER, E. F. Constructivism, conceptual change and science teaching: where are we going? *Investigaçoes em Ensino de Ciências*, p. 20-39. 1996.

- MORTIMER, E. F. *Linguagem e Formaçào de Conceitos no Ensino de Ciências*. Belo Horizonte:

Editora UFMG. 2000.

- MORTIMER, E. F.; AMARAL, L. O. F. The hotter the better: heat and temperature in the teaching of thermochemistry. Revista Quimica Nova na Escola, n. 7, p. 30-34, 1998.
- MORTIMER, E. F, SCOTT, P. & EL-HANI, C. N. "Theoretical and epistemological bases of the conceptual profiles approach." Tecnè, Episteme y Didaxis: TED 30 (2011).
- MORTIMER, E. F.; MACHADO, A. H. Quimica: Ensino Mèdio. 2. ed. Sao Paulo: Scipione, 2013, v.2.
- PERNAMBUCO, M. M. C. A. Educaçao e escola como movimento do ensino de ciências: a transformaçao da escola pùblica. 1994. Thesis (Doctorate in Education) - Faculty of Education, University of Sao Paulo, Sao Paulo, 1994.
- PERUZZO, F. M.; CANTO, E. L. Quimica na abordagem do cotidiano. 4. ed. Sao Paulo: Moderna, 2006. v.2.
- PONTUSCHKA, N. (Org.). Daring in dialog: interdisciplinarity in public schools. Sao Paulo: Loyola, 1993.
- POSNER, G. J., Strike, K. A., Hewson, P. W., & Gerzog, W. A. Accommodation of a scientific conception: Toward a theory of conceptual change. Science Education, p211227. 1982.
- PRETTO, N. Languages and technologies in education. In: Candau, V.M. (Org) Cultura, linguagens e subjetividade no ensinar e aprender. Rio de Janeiro: DP&A, 2ª edition, p. 135160, 2001.
- RAFAEL, F. J. Elaboration and application of a teaching strategy on the concepts of heat and temperature. Dissertation (Master's Degree in Teaching Natural Sciences and Mathematics) - Center for Exact and Earth Sciences, Federal University of Rio Grande do Norte, Natal, 2007.
- RIBEIRO, R. M. L.; MARTINS, I. The potential of narratives as a resource for teaching science: an analysis of physics textbooks. Ciência & Educaçao, Bauru, p. 293309, 2007, v.13.
- ROMMETVEIT, R. On the architecture of intersubjectivity. In R. Rommetveit & R.M. Blakar (Eds.), Studies of language, thought, and verbal communication, London: Academic Press, p. 93-108. 1979.
- SANT'ANNA, B. et al, REIS, H. C.; MARTINI, G.; SPINELLI, W. Conexoes com a Fisica. 1. ed. Sao Paulo: Moderna, 2010, v.2.
- SANTOS, W. L. P.; MOL. G. S. Quimica Cidada. 2. ed. Sao Paulo: AJS, 2013, v.3.
- SCHURMANN, P. F. Luz y Calor. Espasa-Calpe: Buenos Aires, Argentina. 1946.
- SERRA, G. M. D. and ARROIO, A. Analysis of the papers presented at the ENPECs from 1997

to 2005, which addressed the use of microcomputers as a resource for learning. In: Anais do VI ENPEC - Encontro Nacional de Pesquisa em Educaçao em Ciências, Florianópolis, Santa Catarina, 2007.

- SILVA, A. F. G. da. The construction of the curriculum from a critical popular perspective: from meaningful speeches to contextualized practices. 2004. 405 p. Thesis (Doctorate in Education) - Pontifical Catholic University, Sao Paulo, 2004.

- SILVA, D. Estudo das Trajetórias Cognitivas dos Alunos: no ensino da diferenciaçao dos conceitos de calor e temperatura. PhD Thesis, Faculty of Education, USP, Sao Paulo, 1995.

- SILVA, D; NETO, V. F; CARVALHO A. M. P. Teaching the distinction between heat and temperature: a constructivist vision. In: NARDI, R. Current issues in science teaching. Sao Paulo: Escrituras Editora, Chap. 7, P. 61-76. 1988.

- SILVA, E. J. The difficulties encountered by high school students in the concepts of heat and temperature. Monograph (Degree in Physics) - State University of Cearà, Fortaleza, 2007.

- SILVA JR., W. R. R. Development of a program for teaching thermal physics at secondary level. Dissertation (Master's Degree in Teaching Natural Sciences and Mathematics) - Center for Exact and Earth Sciences, Federal University of Rio Grande do Norte, Natal, 2007.

- SOLANGE, C. G. Teacher Training on New Technologies in Education. Monograph (Degree in Biology), Northern Consortium for Distance Education, UNB and UEGO, Brasilia, 2011.

- TONIATO, J.D.; FERREIRA, L.B.; FERRACIOLI, L. Technology in Physics Teaching: A Review of. XVI National Symposium on Physics Teaching. In: X Encontro de Pesquisa em Ensino de Fisica, Londrina-PR, 2006.

- TRIVINOS, A. N. S. Introduçao à pesquisa em ciências sociais: a pesquisa qualitativa em educaçao. Sao Paulo: Atlas, 1987.

- TUNES, E. Scientific concepts and the development of verbal thought. Cadernos CEDES, p. 29-39. 1995.

- VYGOTSKY, L.S. Mind in society: The development of higher psychological processes (M. Cole, V. John-Steiner, S. Scribner & E. Souberman, Eds.). Cambridge, MA: Harvard University Press. 1978.

- VIGOTSKI, L.S. The construction of thought and language (P. Bezerra, transl.). Sao Paulo: Martins Fontes. 2000.

- ZYLBERSZTAJN, A. Spontaneous conceptions in physics: examples in dynamics and implications for teaching. Revista de Ensino de Fisica, vol. 5, n° 2, 1983.

More Books!

yes

I want morebooks!

Buy your books fast and straightforward online - at one of world's fastest growing online book stores! Environmentally sound due to Print-on-Demand technologies.

Buy your books online at

www.morebooks.shop

Kaufen Sie Ihre Bücher schnell und unkompliziert online – auf einer der am schnellsten wachsenden Buchhandelsplattformen weltweit! Dank Print-On-Demand umwelt- und ressourcenschonend produzi ert.

Bücher schneller online kaufen

www.morebooks.shop

info@omniscriptum.com
www.omniscriptum.com

OMNIScriptum

MIX
Papier aus verantwortungsvollen Quellen
Paper from responsible sources
FSC® C105338

FSC
www.fsc.org

Printed by Books on Demand GmbH, Norderstedt / Germany